The Palgrave Macmillan Animal Ethics Series

In recent years, there has been a growing interest in the ethics of our treatment of animals. Philosophers have led the way, and now a range of other scholars have followed from historians to social scientists. From being a marginal issue, animals have become an emerging issue in ethics and in multidisciplinary inquiry. This series will explore the challenges that Animal Ethics poses, both conceptually and practically, to traditional understandings of human-animal relations. Specifically, the Series will:

- provide a range of key introductory and advanced texts that map out ethical positions on animals;
- publish pioneering work written by new, as well as accomplished, scholars;
- produce texts from a variety of disciplines that are multidisciplinary in character or have multidisciplinary relevance.

For further information or to submit a proposal for consideration, please contact Amy Invernizzi, amy.invernizzi@palgrave-usa.com.

Anja Heister

Beyond the North American Model of Wildlife Conservation

From Lethal to Compassionate Conservation

palgrave
macmillan

Anja Heister
Independent Researcher
Missoula, MT, USA

ISSN 2634-6672 ISSN 2634-6680 (electronic)
The Palgrave Macmillan Animal Ethics Series
ISBN 978-3-031-14148-5 ISBN 978-3-031-14149-2 (eBook)
https://doi.org/10.1007/978-3-031-14149-2

Cover illustration: Volodymyr Burdiak / Alamy Stock Photo

This Palgrave Macmillan imprint is published by the registered company Springer Nature Switzerland AG.
The registered company address is: Gewerbestrasse 11, 6330 Cham, Switzerland

I dedicate this book to all earthlings, especially the furbearing ones.

Foreword

Anthropocentrism is the dark side of our species' belief in its own uniqueness. Under this rubric, all other living creatures, lacking our version of language, are unable to demonstrate a sufficiently conscious awareness of self to warrant equal treatment as rational and moral beings. These assignations ensured that non-humans have long been perceived as lesser creatures with no claim to moral standing or even the right to coexist with us as fellow autonomous beings. That these spurious rationalizations reflect a high level of cognitive dissonance, or that morality is rooted in benevolence, not reason, have had little impact on the predatory relationship between humans and wildlife through time, especially in North America where continuous settlement patterns and rapacious resource extraction have put wildlife at threat.

Modern wildlife management under State control originated in North America in the early twentieth century as a component of the Conservation movement which advocated "wise use" practices to conserve natural resources like forests, soils, and minerals. Wildlife was included in this conservation "package" as a species-specific renewable resource to be managed in the human interest. For most of the twentieth century and beyond, official wildlife management practices focussed on game animals while unreservedly embracing Aldo Leopold's utilitarian maxim which equated desirous wildlife to a harvestable crop. Despite modern, more sensitive modifications to human belief systems about the proper treatment of wildlife, this blatant anthropocentric stance has maintained itself in official wildlife policymaking, supported in large part by a well-organized and powerful coterie of vested hunting and trapping interests.

It is to this ecologically untenable and morally deficient mindset that Anja Heister has turned her attention in her well-researched monograph, *Beyond the North American Model of Wildlife Conservation: From Lethal to Compassionate Conservation*. Heister decries the officially sanctioned plunder of fur-bearing and game animals as cruel, unnecessary, morally wrong, and reflective of an anthropocentric belief system. Her call for a more affective relationship with animals based on compassionate conservation grounds is central to her narrative.

The North American Model of Wildlife Conservation is of particular interest to Heister and for good reason. Conceived in the Conservation era and nurtured by Aldo Leopold's beliefs about sustainable game animals, the Model was fashioned at the dawn of the new millennium by a group of interested parties, including the Boone and Crockett Club, and came into being as a set of seven guiding tenets respecting public rights to non-commercial fish and wildlife. Though lacking any legal power or precedent, the Model has become widely accepted by professionals and agencies and has been incorporated into government policymaking in the United States and Canada. Heister recognizes the Model's narrow base despite its well-funded supporters and takes it to task for its heavy endorsement of sports hunting, its overt hypocrisy, its reliance on tired science, and for its species-specific approach which fails to recognize animals as individuals.

This latter point comes through strongly in the narrative. The failure or refusal to consider the value of individual lives is at the root of anthropocentric wildlife policies. The designation of wildlife as a natural resource presupposes its human utility. Only in a belief system that fails to recognize animal consciousness or individual lifestyles that parallel our own in their experiential nature, can an aggregate of living beings be equated to truckloads of coal. Heister acknowledges this aberration often in her narrative and attributes it to several factors.

Though she does not exclude women, Heister's statistics demonstrate a dominant male presence in the hunting and animal exploitation communities as well as in their sponsor and lobby groups. On the other hand, she indicates the heavily disproportionate role of women in animal welfare and advocacy groups. The contrasts are clear enough for her to suggest that women generally are more suited than men to espouse her compassionate conservation ethic.

Heister makes a powerful point in isolating the relatively low numbers of people involved in the hunting and trapping industries in contrast with overall population numbers, the vast majority of whom have no interest in hunting or trapping. Indeed, Heister contends accurately that most people are sympathetic towards wild animals. However, they remain largely unaware of wildlife management philosophy generally, let alone the North American Model of Wildlife Conservation, or the sickening practices of the Safari Club. This explains why animal exploitation groups manage to maintain their position of financial and political privilege. To Heister, most Americans might lack awareness of the barbarities associated with hunting and trapping, but both in terms of numbers and behaviour they are far less anthropocentric than the small influential minority who cleave to the outmoded ideals of the North American Model of Wildlife Conservation.

For Heister, the solution to a meaningful shift away from anthropocentrism lies in the affective domain. To her, empathy is the key. She asks humans to empathize with the plight of violated creatures who cannot speak for themselves. She thus calls for a greater public awareness of the pain, misery, and emotional and physical suffering endured by animals in traps, and less public tolerance for the practice of destroying and ruining the lives of healthy wild animal for sport. It is her hope that empathy will find a collective voice.

Readers will learn a lot in this book. A painful litany of abuse and violence emerges as Heister lays out her facts about hunting and trapping. Readers will also be given hope that this predatory anthropocentric attitude towards wild creatures and especially fur-bearing animals, need not prevail in a modern society where it is neither needed nor wanted, and where empathy supersedes power.

Department of Communication, Media, and Film
University of Calgary,
Calgary, AB, Canada

Max Foran

Prologue

Liberating a Trapped Pine Marten as a Life-Changing Experience

Several winters ago, my partner Dave, his son and I, while cross-country skiing in a national forest outside of Missoula, came across a victim of Montana's furbearer trapping program. A female pine-marten was hanging from a tree, with her front paw caught in a leghold trap; her body motionless, her head tipped to the side, a huge pile of scat on the ground beneath her indicated she had been hanging there for an extended period of time. Upon our approach, she began to twitch with fear. We were shocked, but our natural, empathetic instincts to help kicked in and we started working to free her from the merciless grip of the steel trap.

She then ran away, her gait turning to a limp once she assumed a safe distance from us. All of a sudden she stopped and turned around to look at us as if to thank us. Our eyes met. I will never forget that moment. She became a someone to me. She then disappeared in the woods. The three of us were in tears. If the trapper had found this animal before we did (*if* he had returned to the trap site), this young, tortured, helpless pine marten would have been likely clubbed, stomped to death, strangled, or shot, after having spent an unknown time hanging in the trap—a fate that no animal, whether domesticated or wild, deserves.

Not a day has gone by since meeting that pine marten that I haven't thought of her, and other wild animals out there in the woods, trying to carve out a living for themselves. Did she survive? Did she overcome her horrific trauma caused by a trap that so violently held her in her painful and merciless grip? Did she ever meet a mate, raise a family and continued the life a pine marten is supposed to live?

Acknowledgments

I want to express my deepest gratitude to my closest and dearest friends who are always there for me, and whose friendships I cherish: Petra Menke, Hayley Hesseln, Lee and Stan Gautier, Connie Poten, Deborah Slicer, Robin Dorman, Carla Abrams, and Susan Waters. Heartfelt thanks to my community of terrific friends in Germany and the United States, who have been with me through the many ups and downs. In particular, I thank Carmen Carlisi, Yvonne Fendt, Wendy Bowman, Walter Hopf, Ellen Rappitong and Veronika Weingart, my childhood friends, who continue to welcome me even after long times of absence.

My sincere thanks to Professors James Burchfield, Jeffrey Gritzner, Lee Metzgar, Deborah Slicer and Daniel T. Spencer (Dan) for their steadfast encouragement, insights and constructive feedback throughout many years of my academic studies.

I am so grateful to folks who work tirelessly to help nonhuman animals whether in the field, in the classroom, or in the solitude of the writing experience or otherwise, including Max Foran, Eduardo Gonçalves, Barbara Stagno, Jen Nitz, Rob Domenech, Brooke Vogtsberger-Tanner, Brooks Fahy, Sally Mackler, Jim Robertson, Anne and Peter Muller, John Eberhart, Lane Ferrante, Lisa Robertson, Helen Kopnina, Eileen Crist, Joe Gray, Paul Cryer, Simon Leadbeater, Dave Pauli, Stephen Capra, Jim Schmitt, Joy Stevens, Clare Beelman, Marc Bekoff, Camilla Fox, Arian

Wallach, Adrian Treves, Rocky Sehnert, and so many more folks deeply committed to our efforts to improve our relations with other animals.

Thank you to Mowgli, Olivia, and Lena, beloved canine family members, whose company provides wonderful times of shared comfort, joy, friendship, and laughter.

And finally, thanks to David Taylor, the foundation of my sanity and spirit through all these years. I really look forward to what is yet to come, you know.

Contents

1

Introduction

The Fate of Wolf O-Six and Family

Some of us are destined to become a leader and celebrity. She was one of them. She had left her home as a young adult and spent the following years living independently and sometimes joining others. She did things in her life regardless of convention, even her choice of a mate, or mates rather, was out of the ordinary. None of the strong males persistently courting her along the way had impressed her sufficiently enough to settle down with. Instead, when she finally did, upon reaching middle-age, she chose two younger brothers (M754 and M755) to become the fathers of her children and to spend the rest of her life with. Nicknamed "O-Six" for the year she was born,[1] *she quickly rose to being a leader and a celebrity. She was loved and admired by her large crowd of human fans, who, every year, flocked by the tens of thousands to Yellowstone National Park, America's first national park established in 1872 to observe her and her family. O-Six was the leader of the highly visible and successful Lamar Canyon wolf family (pack) and known for her extraordinary confidence, loyal character, dazzling appearance, including a strong physique, she was a devoted companion to her two mates and a fiercely protective mother to*

A. Heister, *Beyond the North American Model of Wildlife Conservation*, The Palgrave Macmillan Animal Ethics Series, https://doi.org/10.1007/978-3-031-14149-2_1

her children. While it usually takes the cooperation of several family members working together to take down and kill an elk, who can weigh around 700 pounds (eight times as heavy as she), O-Six would do this on her own to provide food for her family.

Had her significant others, M754 and M755 been humans, they would have been described as "great guys"; They would have been married, loyal husbands who also enjoyed spending a lot of quality time with their children. M754 was good-natured, and despite his larger size, he didn't mind his brother being the male Alpha. Motherhood for O-Six started in 2010 and continued throughout 2012, the year when the family had four pups. The Lamar Canyon Valley wolf family was thriving. Life was good.

Then, tragedy struck twice during the last two months of the year 2012. On November 13, the entire family was on the move eastward within their territory but unbeknownst to them, they had crossed the invisible lines of Yellowstone when M754 was shot by a trophy hunter. Immediately after the loud noise of the gunshot, the wolf family members scattered into all directions. They got back together again later and returned to Yellowstone perhaps not yet realizing that M754 was missing. Once they got back to their main den and rendezvous site, they must have realized that one of them was missing, repeated howling ensued, typical behavior of a wolf family letting missing members know where they are. It was in vain.

About three weeks later, the wolf family moved eastwards again and were close to the area where M754 had been killed, probably still looking for him. They were unaware of the consequences of leaving the park—marked by invisible lines in the landscape only known to humans—that had protected them all these years from 'sportsmen.' That day, December 6, 2012, O-Six would share the same fate as M754, when she was shot by a trophy hunter.

Her death left her human fans, scientists, and wolf enthusiasts alike devastated all across the world. The New York Times even published an obituary. These human tributes to O-Six's life could however not ease the devastating blow to her family now stricken with grief and chaos. Unable to cope with the loss of O-Six as their leader, her family soon fell apart. Some of the family members moved to the east and outside of Yellowstone into Wyoming, other family members stayed with M755 but because there were no females unrelated to him, he had to leave the family and become a lone wolf in search of a mate. The once resilient and socially coherent family broke up into three different, smaller groups.[2]

O-Six and M754 were just two of hundreds of millions of wild animals killed by hunters and trappers primarily for recreation, fun, trophies and private profit, every year in the United States alone. The state wildlife agency in Wyoming where O-Six and M754 were shot, sells a license to shoot a wolf for $21 and in most of the state wolves can be shot on sight; their deaths marked as a mere statistic on that year's 'harvest report,' that tracks the sportsmen's 'success' in 'harvesting' wolves and other wild animals. The impacts on the animals' mates and family, including dependent young, is of no interest—after all, according to traditional wildlife management, wolves and other wild animals do not matter as individuals—instead, what matters is the species and their continuation into the future for the primary purpose of recreational hunting and trapping. Unlike humans, wild animals are considered 'things,' and unlike wild animals, humans are 'persons' with rights. Canada-based author and Professor Emeritus of Communication, Media and Film, Max Foran explains:

> It is because of this anthropocentric—human entitlement—belief system that is endemic in wildlife policies and laws that predefines "an ongoing downward trajectory despite remedial interventions and public advocacy. Whether as a resource to be exploited or as a sustainable component of an ecosystem, wildlife are victims in management regimes already constrained by a wide variety of anthropogenic disturbances. Since wild creatures receive minimal ethical consideration and since human incursions into their domains are ongoing and relentless, the management outcomes become predictable.[3]

Indeed, for nearly 150 years, the view of wild animals as 'renewable natural resources' and 'property' to be managed, controlled, and used has dominated wildlife management and conservation in the US. The so-called 'North American Model of Wildlife Conservation' (hereinafter referred to as NAM, or the North American Model) is the driver of this strong anthropocentric and utilitarian stance, which is not only promoting a perpetual killing spree of wild animals nationwide but has also resulted in a staggering spiral of plant and animal extinctions globally.

What the Book Is About and Why It Is Needed

The focus of this book is on the NAM, with the aim of exposing the anthropocentrism, the belief in human superiority and entitlement that undergirds traditional wildlife management with hunting and trapping as its cornerstones. The NAM represents this state-government-wildlife-management-system, which fails to consider the intrinsic value of individual animals, worthy of respect and with flourishing lives, and is thus the cause for much suffering and death of wild animals, their mates, and families.

Most people believe that government agencies are involved in conservation to protect wildlife. Learning that the principal activity of most state wildlife agencies is the administration of hunting, fishing and trapping programs, sale of kill licenses and the preparation and enforcement of regulating these activities comes as a shock to many. The general public is also largely unaware of the NAM's existence and "its influence on the biological and political landscape."[4] It has been estimated that hunters in the US alone kill between 100 and 200 million wild animals annually, the majority for 'recreation.'[5] In addition, trappers kill between 6 and 21 million wild fur-bearing animals annually.[6] Furthermore, the United States Department of Agriculture's Wildlife Services agency—a body that one documentary rightly exposed as an "unaccountable, out-of-control, wildlife killing machine"[7] relies heavily on leghold traps and strangulation snares among other indiscriminate devices in their annual slaughter of millions of wild animals under the guise of livestock protection.

Hunting and trapping of wild animals not only occur on the vast majority of public lands but also on more than half of the areas within the National Wildlife Refuge System in the United States which encompasses more than 95 million acres and includes 568 national wildlife refuges. Managed by the U.S. Fish and Wildlife Service (USFWS), these refuges harbor more than 700 species of birds, 220 species of mammals, 250 reptile and amphibian species, and more than 1000 species of fish. The word 'refuge' derives from the French word "to flee" and thus means a safe place to flee from or flee to. In human affairs, the common usage of the word refuge means a place safe from harm. This is not the case when it comes to wild animals—they may be refugees, but they will not find a place to rest and one that is safe from human harm, specifically from sportsmen.

Hunting is allowed on 340 wildlife refuges and 37 wetland management districts. On its various website pages dedicated to the promotion of hunting in wildlife refuges, USFWS explains that the refuge habitat produces a surplus of renewable resources that are harvestable. In the agency's view, hunting is a healthy recreational activity that is not only steeped in tradition as part of America's heritage but does not threaten the resident animal populations or species. Motivated by such convictions, the USFWS is involved in constant efforts to expand hunting opportunities for sportsmen by increasing acreage, the number of species to shoot and the types of weapons that can be used.[8] The agency even advertises wildlife refuges as destinations for "trophy hunting … or bucket-list destination hunting."[9] Finally, it states that not wildlife watching but hunting is seen as the priority use of wildlife refuges, despite the fact that of the 59.7 million annual visits, only 2.4 million were for hunting compared to 42.1 million people, who visited to watch birds and observe wildlife.[10]

Information about trapping is more difficult to obtain, but a survey conducted by the USFWS between 1992 and 1996 revealed that 280 of the 517 National Wildlife Refuge units had trapping programs, and the USFWS website provides a long list of lists of refuges where trapping occurs (USFWS n.d.)[11] And if you thought that this was already over the top, you would be wrong: Under the headline "Unforgettable Hunting on Wildlife Refuges," the USFWS provides a hunting guide that states that hunting is allowed in all of Alaska's 16 wildlife refuges.[12] According to the U.S. Department of the Interior website, " … there are 76 areas managed by the National Park Service, 336 national wildlife refuges and 36 wetland management districts managed by the U.S. Fish and Wildlife Service, and over 220 million acres of BLM-managed public lands—in addition to most Bureau of Reclamation lands—that allow hunting in accordance with federal and state regulations and laws" (DOI 2017).[13] Many of these refuges are simply waterfowl production units for waterfowl hunters, and thus beavers, muskrats, raccoons, skunks and other animals that eat bird eggs and change water levels are routinely trapped and killed. The term "wildlife refuge" is therefore highly misleading, and yet another strategy to disguise the violent killing of wild animals by 'sportsmen' and promoted by governmental agencies alike. Far from being places where 'game' and 'furbearing' animals can find refuge, these are areas that produce, according to Also Leopold's early thinking, an overflow of game animals that can be shot outside, and today, within these areas as well.

In direct opposition to the human supremacy philosophy that wildlife is a resource for human use, some wildlife conservation organizations are exposing the exploitation of so-called refuges, believing that the general public will not stand for it. For example, Born Free USA, a leader in animal welfare and wildlife conservation, launched a new campaign, #TrapFreeTrails in 2020, which calls for an end to trapping in the National Wildlife Refuge System.

The insistence by NAM promoters of the right to kill wild animals for frivolous reasons and the support of such activities by government agencies against the protection of wild animals should not be tolerated. Through its promotion of hunting and trapping, the NAM promotes negative human behaviors such as aggression, cruelty and violence against wild animals, which cause great suffering and harm to sentient animals. These negative traits are detrimental to society's overall efforts to increase pro-social behaviors such as respect, love, empathy and compassion, as well as finding connections to the 'more-than-human-world.' It is an undeniable fact that human-caused destruction of the nonhuman world has reached catastrophic levels. Indeed, under "human supremacy" as leading scholar and author, Eileen Crist (2017) calls the most virulent strand of anthropocentrism, the human relationship with wild animals, is largely one of violence, power and control that treats fellow beings and nature as commodities. Human supremacy is "the collective, lived belief system that humans are superior to all other lifeforms and entitled to use them and their places of livelihood."[14] Such a lived anthropocentric worldview blocks empathy, makes us blind and apathetic to the pain, suffering and death of other beings, and an awareness of the mutual interdependence nearly impossible; this must change.

With a history that reaches back to the turn of the 19th century, sportsmen have created a wildlife management system that today, is highly guarded against the interests of the majority of Americans, who neither hunt nor trap. Termed the "Conservation-Industrial Complex,"[15] this system goes well beyond state and federal wildlife agencies and includes pro-hunting, pro-trapping industries and gun rights associations, each worth billions of US dollars.

These entities have been the main decision-makers of wildlife policy and they together, continue to ensure that wild animals remain subjugated and disempowered, and public influence limited, while collectively providing significant organizational, financial, and political resources in support of sportsmen's interests. With that, they maintain a value system that perpetuates the dichotomy between humans and nonhumans that insists on the inferior status of the latter and promotes and protects the 'sustainable' killing of wild animals for such frivolous reasons as recreation, sport, commerce, food and trophies.

The fates of hundreds of millions of wild animals are in the iron grip of, and decided by, this conglomerate of 'sportsmen' that has but one goal: to maintain control over wild animals and their habitats for the continuation of hunting, angling, and trapping. Predictably, this pro-hunting and trapping industry comes out in full swing whenever the public attempts to curtail recreational hunting and trapping or governmental lethal management of predators.

There is a noticeable value shift among the American public toward recognizing the rights of animals, with an increase in mutualism and empathy rising for individual wild animals. However, these positive changes for nonhuman animals and human society alike are deemed as threats to the survival of the NAM and, consequently, sportsmen and wildlife agencies are doubling down in their efforts to recruit new hunters and to convince the public of the importance of hunting and trapping, along with the claimed indispensable role of sportsmen for conservation.

In the process, the NAM has aggressively been promoted in the last two decades through different media and in interactions with the public. In 2002, the National Association of Fish and Wildlife Agencies (AFWA), representing all 50 U.S. state wildlife agencies in North America,[16] formally endorsed the NAM as the officially accepted paradigm for federal and state wildlife agencies in the United States and Canada (Mexico is excluded). Successful nationwide outreach programs have invested heavily in recruiting, retaining and reactivating efforts (R3) with every state now having an R3 representative while hunting organizations hide behind the label of 'conservation.' The NAM is being promoted strategically in interactions with the public, with relentless and false memes that portray 'hunting and trapping as conservation;' in fact, as necessary and even beneficial for wildlife species.

The argument presented in this book is that we must change the paradigm from exploitation to protection and peaceful co-existence with wild animals. We need an approach to conservation that de-centers humans, recognizes the intrinsic value of individual wild animals, and promotes ecosystem-thinking of which humans are a part.

The convergence of increased concerns by conservationists outside of traditional wildlife management and growing sectors of the public alike about the wanton killing and casual cruelties—crimes against wild animals and their families—has led to the call for a much-needed, new approach in our relationship with wild animals —'Compassionate Conservation,' a growing international movement that brings empathy into the decision-making process that affects wild animals. Compassionate Conservation is responding to the increasing sense of empathy for individual nonhuman animals within the general public, which no longer agrees with the paradigm of wanton killing of wild animals for recreation, sport, trophies and commerce. With a focus on empathy and compassion and a liberating rejection of anthropocentric views, each of us, conservationist or not, can adopt the ethical view that all animals deserve lives of respect and dignity, and thus make changes toward an inclusive, and non-violent lifestyle.

The efforts of promoting compassion for wild animals, locally, as well as globally, are linked to other social movements that seek to legally protect nonhuman animals through the concepts of rights, personhood, and property owners. The survival of wild animals and their wellbeing are dependent upon our belief system, especially at this pivotal moment, when faced with the intensifying threats of an explosive human population, climate change, economic development, urban sprawl, and an accelerating rate of habitat loss.

What it comes down to is as simple as this: if we continue to insist on human supremacy, fellow animals (domestic and wild) will continue to suffer and die; only if we change our belief system, will they have a chance to survive. Everything is at stake for them.

Chapter Overview

As part of the lens through which this book critiques the North American Model of Wildlife Conservation (NAM), Chap. 2 begins with the animal standpoint theory. Considering the nonhumans' points of view helps to illuminate the NAM as a social construct that allocates all power and control over wild animals' fate to humans. As such it represents a framework of oppression that keeps animals—wild and domestic—disenfranchised and subjugated. At the same time, the animal standpoint is crucial as it is a non-anthropocentric approach that holds the potential to help us overcome anthropocentrism. This chapter discusses the philosophical ideas behind the animal standpoint and addresses the changing contemporary perspectives among growing sector of the American public on wildlife that has the potential to shift our relations toward egalitarianism. The chapter ends with an urgent proposal to recognize wild animals as fully conscious beings, who live and thrive in their world competently and skillfully, possessing their own species-specific characteristics while also sharing many essential traits with humans. Yet, they are being victimized by the actions of sportsmen, governmental agencies and other parts of the Conservation-Industrial Complex. A particular vision for social justice moments would include the representation and protection of wild animals in our human-centric process of democracy, to finally achieve an end to our war against other animals (see eco-democracy in Chap. 8).

Chapter 3 tracks the historic development of the North American Wildlife Conservation Model's evolution under the belief of human supremacy. Here, the leading role of (trophy) hunting organizations, specifically the Boone and Crockett Club, emerges in the development of the NAM, along with its leading role in creating governmental agencies focused on wildlife management. The chapter also exposes the

Conservation-Industrial Complex, and its use of propaganda to promote the social acceptability of NAM practices. A closer look reveals this wildlife killing industry as a wealthy and influential network of supporters spanning governmental agencies, organizations, and corporations, resulting in the pervasive, well-funded, legislative defense of sportsmen's interests and killing methods over concerns for the lives of individual wild animals.

Chapter 4 presents the current critique of the NAM by the conservationist community, for example, its selective focus on 'sustainable use' of certain animal species while ignoring others, and its dictate of aggressive persecution of predators, the exclusion of the general public and nonhunting conservationists. The chapter's discussion scrutinizes certain claims that equate hunting and trapping with, and as indispensable for conservation.

Chapter 5 investigates how the NAM's belief in human entitlement plays out in their selective use of ethics to facilitate the exploitation of wild animals. Two topics are central here, which are how the NAM 1) only looks at the whole (animal species and population, ecosystem) and 2) how the NAM ontologizes wild animals as 'natural renewable resources,' as objects and human property (public, government and private) to be (lethally) used. The chapter demonstrates how the NAM fails to properly consider individual wild animals as part of the moral community. The chapter's discussion addresses the NAM's inherent dualistic worldview that dangerously perpetuates the human-nonhuman divide, which, along with other insufficiencies, renders the NAM's ethics as almost willfully opposed to curbing society's ongoing destructive behavior toward the more-than-human world.

Chapter 6 provides a brief overview of two scientific concepts used in the service of hunting and trapping—the 'harvest theory' and adaptive (harvest) management as examples of the type of Cartesian science that informs NAM's policies. The chapter also presents a discussion of some of the main arguments made by proponents to justify trapping. The final section of this chapter showcases the NAM as incapable of preventing extremists among its own ranks, who want to persecute wild animals, especially wolves, even further without any restraints at all. In an eerie repeat of history, it is the persistent hostility against predators by state wildlife agencies, legislators, ranchers and sportsmen that is driving a massive wolf slaughter taking place in Western states in and beyond Montana.

Who is monitoring how sportsmen interact with wild animals in the woods? Chapter 7 addresses this question by providing some examples of cruel activities that sportsmen are involved in and support. From a discussion of spring bear hunts to wildlife killing contests and recreational and commercial trapping, this chapter explores the cruelty and violence inherent in sportsmen's activities and provides an in-depth look at trapping as animal cruelty. The intent here is to raise awareness of hunting and trapping as acts of human aggression—crimes against unsuspecting and defenseless wild animals—and of the intersectionality of violence.

Chapter 8 opens with a story told to me by a former hunter, which sets up the stage for delving into the importance of positive emotions in general, and the value of empathy and compassion specifically in our relations with nonhuman animals. The chapter further explores the concept of Compassionate Conservation as an approach to conservation that promotes the wellbeing of individuals, populations, and ecosystems with its leading principle, *First, do no harm!* This chapter then presents insights from several forward-thinking authors, including those who argue for an expansion of our legal system to recognize individual nonhumans as legal persons with rights, and wild animals as property owners. Other scholars are exploring a path toward Ecodemocracy where wild animals would be assigned stakeholder status and voting rights in policy development via human proxies; these humans would need to be specially trained, for example through the "Council of All Beings," a process to learn how to step aside from our human identity and speak on behalf of another lifeform. The chapter concludes that we need a life-affirming approach to conservation where every individual animal counts and is recognized as a living being with intrinsic value. Under the triple-threats of human population growth, plant and animal species extinctions and climate change, a paradigm change of our culturally entrenched and ubiquitous human supremacy worldview is urgent and needed more than ever. This chapter concludes with suggestions for actions everyone can take to help nonhuman animals, one's own health, and the planet. Emphasis is on the need to end the killing of wild animals for frivolous reasons and advance the protection of nonhuman animals as a social justice issue. Changing our ways of being by adopting a plant-based diet as an expression of a non-violent lifestyle emerges as a priority and a powerful way for us to resist and transform the exploitative and destructive corporate world we all live in.

Definition of Terms

From a biological perspective, human and nonhuman mammals are both animals, but I distinguish between them for this project as humans and nonhuman animals for practical reasons. By using *wildlife* in this book, I follow the traditional understanding of the term to include birds and mammals when discussing traditional wildlife management, trapping, hunting and conservation. When I present my own views, however, I use the terms *wild animals*, *wild nonhuman animals* or wild individuals to express my disagreement with this aggregate term that lumps varied animal species into one term. I use *who* and not *that* when describing nonhuman animals because they are not objects but subjects, and I consider each species as a nation or as *peoples*. I also use plural when I refer to specific species; thus, I do not write about *the red fox*, but rather *red foxes*. The term *harvest* is routinely used by the wildlife profession in lieu of the term *killing* of animals. One cannot 'harvest' a fellow animal as harvest implies that work has gone into what eventually will be harvested but the animal has done that work her/himself and no 'sportsman' has helped with raising a wild animal. Most importantly, the term is exemplary for the human supremacy undercurrent of wildlife management that apparently sees no difference between an intelligent, sentient and independent wild animal and a bushel of corn. Despite my objection though, I use the term when referring to NAM proponents' ideas and concepts in order to avoid confusion and use killing or *exploitation* whenever possible. Finally, the term *furbearer* has a clear commercial and commodity connotation, describing animals that have traditionally been killed for their fur.[17] I maintain using the term to distinguish this grouping of animals from other wild animals who are exploited in other ways. The term *game* animals should not be used as it conveys the use of these animals and implies fun and play when in reality it is activities of deception (luring wild animals in front of guns, arrows and into baited traps) that result in the death of the animal. However, the term is so pervasive in wildlife management that I do not see a way around it when I describe hunters' activities. I use the term the *NAM* in order to convey ideas, concepts and views of its proponents, not the model itself. Kheel (1996) has deconstructed the claim of hunting being a type of 'sport,' thus, the term 'sportsman' is yet another euphemism to conceal the killing of wild animals by humans. Regrettably, for a lack of a better term and for a better flow for the reader, I do use the term 'sportsmen' as an umbrella term for hunters and trappers—sport, recreation, trophy and profit—inclusive of women and all other gender identifications. '*NAM promoters*' is used here to describe primarily a small but highly influential group of wildlife 'professionals' and managers along with representatives of hunting, trapping and pro-gun organizations, academics at universities and all other pro-NAM supporters.

This book focuses on the trapping of wild animals, and only occasionally ventures into the issue of hunting to keep the chapters less cumbersome but the reader should hear trapping *and* hunting in their mind whenever one or the other term is mentioned. In describing our 'relationship' with other animals I use "human-animal relations" with other (wild) animals and only sometimes, and reluctantly, "the human-nonhuman relationship because I believe our treatment of other animals cannot be described as a relationship so long as there is no respect, solidarity and love for the other beings. Furthermore, it seems clear that wild and free animals need a "let them be" and "support them" attitude from us humans, while we want and need to be in a relationship with them. We feel the need to connect with wild animals but I do not think that they do. It's not reciprocal and thus, not a relationship.

This book is motivated by a deep concern for the perhaps hundreds of millions of wild animals, who suffer and whose lives are frivolously and mercilessly extinguished by sportsmen, and governmental officials. It is my hope that this book will make people feel motivated and inspired to speak up for wild animals as subjects with needs, interests and rights, and demand policy changes toward their radical protection. Current efforts toward this goal, including mobilizing for rights of nonhuman animals, nature, and Compassionate Conservation underscore the need to develop a new framework, that is grounded not in exploitation and deaths for other beings but instead motivated by emotions such as empathy and compassion. In keeping with dualistic traditions, animals have been disassociated into discrete pieces within human society and are used in conservation as props, targets and objects of commodities of fur and meat (food)—providing recreational entertainment and personal profit. By bringing attention to the issue of killing wild animals for recreation, trophies, damage control and commerce, the present work is an attempt to help us break away from an old, outdated and pathological paradigm of dominion and difference.

Let's put ourselves in the world of wild animals where they survive by their wit and courage, in order to present their view in a way other than the one dictated by speciesism. I am hopeful that this book helps the reader to recognize killing wild animals for reasons absolved by the NAM as what it is: human aggression and violence against innocent and unsuspecting animals that cause them suffering and their premature and artificial death, for which there is no justification. This book exposes the North American Model of Wildlife Conservation and its violence against wild animals to help us find a better way of thinking and feeling about our fellow animal beings.

Notes

1. For 2006, the year she was born and to distinguish her from her identical sister, who was born one year later in 2007. Researchers called her 832F (this was also her collar number).
2. If neither of the two wolves had been shot, the entire family would have likely returned to Lamar Valley and successfully raised one or two more litters of pups through 2014.

   After the shootings of O-Six and her mate, wolf visibility in Lamar Valley went down significantly in 2013 and 2014. Specifically, the average number of wolves seen decreased 44.3 percent, while the difference in days on which no wolves could be seen at all increased 200 percent during 2013 and 2014 compared to the two years prior to their shootings. Thus, for tourists coming to YNP perhaps for a once-in-a-lifetime experience of seeing wolves, the wolf hunt/shootings reduced that chance for them by 50 percent.

I am very grateful to naturalist, wolf expert and author, Rick McIntyre of Yellowstone, Montana, who shared the details of O-Six's and M754's stories with me.

Note: See this website with a photo of wolf 21 and his true love, wolf 42: Living on Earth: The Reign of Wolf 21 (loe.org).

3. Foran, *Subjugation,* 30.
4. Clark and Milloy, "North American Model," 352.
5. Bekoff and Pierce, *Wild Justice.*
6. White et al., "New Guidelines," 68.
7. Predator Defense, "Bobcat Rescued."
8. For example, see: News Releases - U.S. Fish and Wildlife Service (fws.gov).
9. Unforgettable Hunting on Wildlife Refuges | U.S. Fish and Wildlife Service (fws.gov).
10. Great Recreation | U.S. Fish and Wildlife Service (fws.gov).
11. https://www.fws.gov/refuges/wildlife-conservation/trapping.html
12. https://www.fws.gov/refuges/hunting/unforgettable-hunting-on-wildlife-refuges/
13. U.S. Department of the Interior (DOI), "Everything You Need to Know."
14. Crist, *Abundant*, 3.
15. Serfass et al. "Empowerment and Exclusivity," 108.
16. AFWA, "North American Model."
17. The classification of furbearers in North America comprises both carnivores and rodents, and although the legal definition often varies by state, furbearers mainly include beavers, bobcats, coyotes, fishers, foxes, lynx, martens, minks, muskrats, river otters, skunks, weasels, and wolverines. Thus, in the traditional sense, these are animals who, like things, bear fur and are valued and killed by the trapping community only for their body (similar to trophy hunting, where animals are killed for their antlers, horns, or their teeth (ivory)). From a biological perspective, while all mammals bear fur of some type, only about 100 species worldwide are subject to commercial harvest. Of these, 30 species reside in North

America. Adding to the confusing nomenclature is the fact that in some states' jurisdictions, many species of furbearing mammals, for example, red foxes (*Vulpus vulpes*), badgers (*Taxidea taxus*), coyotes (*Canis latrans*), weasels (*Mustela spp.*), and raccoons (*Procyon lotor*) are legally classified not as furbearers, but as predators, nongame wildlife, or other designations. Finally, several species present in the western United States who originally fit the general definition of furbearer, such as grizzly bears (*Ursus arctos*), wolves (*Canis lupus*), and mountain lions (*Felis concolor*), are now variously managed as threatened or endangered species, or as game or trophy animals.

2

The Animal Standpoint

Introduction: From Marxism to Ecofeminism and Critical Animal Studies

Grounded in the Hegelian master-slave dialectic, Karl Marx and Fredrich Engles developed a standpoint theory of the proletariat in order to pursue a full critical analysis of social class relations in capitalism. Standpoint theory re-emerged as part of 20th-century feminism to expose patriarchal domination and its oppressive power relations held in place by white, male dominator culture in society. From here, in her 2006 discussion of dialogical ethics—engaging in conversations with other animals—American scholar of comparative literature, feminism, and animal ethics, Josephine Donovan, suggested extending standpoint theory to animals.[1]

Standpoint theory recognizes that human society is structured by power relations that produce unequal social locations, whereby one location is occupied by members of the dominant group, and other locations are inhabited by members of subordinate groups. Recognizing that social and political forces generate these unjust power relations, feminist standpoint theory thus rejects the privileging of men and men's interests and the associated devaluing, marginalizing, and harming of women and their interests. Similarly, animal standpoint theory recognizes that the

A. Heister, *Beyond the North American Model of Wildlife Conservation*, The Palgrave Macmillan Animal Ethics Series, https://doi.org/10.1007/978-3-031-14149-2_2

belief in human supremacy and speciesism produces unjust power relations and thus rejects the privileging of humans while devaluing, marginalizing, and harming nonhuman animals. One of the important advantages of standpoint theory is that by "studying subordinated locations and the knowledge they foster not only provides insights into the lives of members of subordinated groups but also casts light on dominant group practices, especially when they reproduce inequality."[2] Understanding this in a wider ecological context, Steven Best has stated that, just like (eco)feminism and postcolonial critical race theories can illuminate colonial, slavery, and racist pathology, so can animal standpoint theory shine a light on human speciesism and unjust domination over nonhuman animals, along with the disastrous social and ecological consequences of our alienation from the natural world—"the pathological humanist project to '"dominate" and "master"' it."[3]

The last four decades have seen an extraordinary increase in public as well as an academic interest in animal lives and human-animal relations, captured by the academic discipline of animal studies. This interest has spread to other disciplines like history, anthropology, sociology, cultural and gender studies, literature and law as well as interdisciplinary and gender studies. However, while the field of animal studies was initially intended to be non-anthropocentric, the discipline has been perceived by some scholars as being co-opted by mainstream human-centric interests again (e.g., the adding of hunting-related classes at colleges and universities), a development that triggered the creation of the academic discipline of critical animal studies (CAS) to emerge, along with its animal standpoint perspective. Influenced by critical theory from the Frankfurt School[4] and feminism, among other sources, critical animal studies "promotes ethical reflection on the way humans treat nonhumans, and to do so from a perspective that is grounded in intersectionality, solidarity and critique of capitalism." The Institute of Critical Animal Studies (ICAS) further specifies that "critical animal studies is rooted in total animal liberation, is explicitly committed to a deconstruction of the binary opposition between human and nonhuman animals and aims "to dismantle all systems of domination and oppression, in hopes for a just, equitable, inclusive, and peaceful world."[5]

Animal standpoint theory examines the origins and development of human societies through the dynamic, symbiotic interrelationships between human and nonhuman animals. It therefore interprets history not from a position that reifies human agency as the sole and autonomous force of history, but rather from a co-evolutionary perspective that sees nonhuman animals as inseparably embedded in human history and as dynamic agents in their own rights. If history is examined "from the animal standpoint, that is, from the crucial role that animals have played in human evolution and the consequences of human domination of nonhuman animals, we can glean new and invaluable insights into psychological, social, historical, and ecological phenomena, problems, and crises." The central thesis of animal standpoint theory, according to Best, is that "animals have been key determining forces of human psychology, social life, and history overall," and that in fundamental ways, "the domination of humans over non-humans underpins the domination of humans over one another and over the natural world."[6] Furthermore, the animal standpoint is essential for a proper understanding of power relations in society, that is to rethink the dysfunctional order that structures our relations to one another, other species and the world. Here, Best points to speciesism and identifies the beginning of animal agriculture as the first form of hierarchy and domination that laid the groundwork for other forms of oppression, power and violence.

Throughout this book, I maintain that taking the perspective and attitude of animal standpoint(s) is essential to overall efforts to 1) de-center humans and move other animals from the periphery into the center of our moral concern, and 2) to dismantle the hierarchy in human and nonhuman animal relations in order to recognize and afford animals their subjectivity (see the unjust power relations exerted by the 'Conservation-Industrial Complex' as examined in Chaps. 3 and 5; for my argument regarding the need for empathy to overcome anthropocentrism, see Chap. 8). From the claim of standpoint theory that "the subjectivity of historically marginalized groups can be properly recuperated only when analyses proceed from their perspective,"[7] follows that wildlife conservation can accurately depict wild animals only when they are conceived as subjects in their own rights and not as objects, things, property, and targets for human's assumed right to use for 'recreation.'

The distinction between object and subject rises to the level of great importance, a topic that Canada-based Professor and animal legal scholar, Maneesha Deckha has addressed: To be recognized as a subject with subjectivity is to be seen as a person "entitled to respect, dignity, autonomy and all the other normal markers of personhood status in Western societies. When treatment that accords with these values does not materialize, injustices arise, most often along the lines of objectification and dehumanization."[8] The important view and the concept of (wild) animals as persons will be further discussed in Chap. 8.

I should note at this point that this book primarily focuses on the pain and suffering of wild animals because they are the most studied emotion due to their implications for animal welfare, well-being and efforts for their liberation. Animals' subjectivity is of course much more complex and involves a wide range of positive emotions and experiences such as joy, curiosity, playfulness, contentment, and excitement. In her essay that deeply explores "Joy," Montana-based poet and philosopher Deborah Slicer adds the following important considerations for a more complex subjectivity of other animals. First, and while fully acknowledging our brutal treatment of other animals, Slicer emphasizes that by remembering the animals' joy, "or, acknowledging the billions of animals who are deprived of it, at least their capacity for joy, other animals become fuller subjects and are less likely to be ontologized as victims." Slicer views the suffering of animals not as being passive but instead, as a form of resistance to our unethical treatment. In reference to the Buddhist concept that 'resisting means to suffer,' Slicer writes, "Other animals do suffer *because* they resist." Related, when we feel joy, the effect is that we want more of it and share it with others, humans and nonhumans all included, because we value it for its own sake. This is perhaps, as Slicer contemplates because joy is something even more fundamental than beauty or justice. In the end, "joy is decentering, and as we pay better attention, love, the less-preoccupied we are.[9]"

I am in full agreement with Slicer, thus it is worth stressing here that domestic and wild animals only become victims of human oppression due to our actions that impact their lives and cause additional and unnecessary pain, suffering, and death. Regrettably, because of sportsmen's actions that detrimentally impact animals, and because pain, suffering and death are the aspects of sentience with the most obvious ethical consequences, these are the issues discussed in this book. I use animal standpoint as comprising multiple animal standpoints. For example, a red fox surely experiences the world very differently than a polar bear, or a bald eagle, yet they have a shared interest in remaining unharmed and staying alive, while they may all feel the pressure and threats of human domination and the strains we put on them.

Hard Realities

From predators such as wolves who keep other animal populations in check, to beavers whose work is indispensable for water retention to pollinators, such as bees all play critical roles in diversifying and stabilizing ecosystems, while maintaining the health and integrity of ecosystems (their homes), on which all life depends. Foran (2018) writes:

> Wild animals are the essence of this interconnected web of life. Wildlife are ecological engines contributing immeasurably to the vital cycles and balances that sustain all life. From soil aeration to nutrient cycling, water purification, pollination, seed germination, and seed dispersal, wild creatures contribute to biotic communities in ways humans have lost or never had. Aesthetically, wildlife surpass any human creation. They present palettes of colour, beauty, and intricacy of form, harmonies of sound, and wondrous balletic, artistic, and athletic displays by being who they are. Entrances to their worlds are invitations to mystery, excitement, apprehension, and above all, the wondrous diversity of life … Above all, they live independently in their own worlds and in their own ways…[10]

Wild animals also work very hard to stay alive. Most females, once they enter the time of pregnancy and motherhood, are burdened with the additional reproduction work, care of their children and having to pay additional attention to dangers and risks threatening not only themselves but their little ones.

Many animals have a short lifespan to begin with. Wild animals already suffer from the impoverishment of their lives associated with unprecedented human population growth, causing suffering to individual animals through increasing displacement, stress, and starvation, and reducing security.[11] They have to cope with a multitude of intra-and inter-species challenges and natural harm, in addition to disturbances caused by humans such as urban sprawl and increased encroachment into their habitat. Other challenges include thirst and hunger, food scarcity, diseases, fear of predators, competition, and psychological stress, injuries due to accidents, conflicts, and psychological suffering, grief, extreme weather conditions including hot and cold temperatures, climate change and associated wildfires, storms and drought, pollution, noise, fragmentation and loss of habitat and thus loss of homes (dens, nests). Given these few examples of a much longer list of challenges, it is easy to understand that wild animals endure a wide array of forms of natural harms that make life in nature far less idyllic than we generally like to believe[12] and to imagine what Spanish philosopher and animal activist, Oscar Horta concludes, that the odds are often against wild animals to the point where "suffering and premature death prevail over positive well-being in nature."[13]

At a minimum, the constant pursuit of wild animals by hunters and trappers, in addition to USDA Wildlife Service agents always on the lookout for wild animals to poison, shoot and trap (in particular, predators such as wolves, coyotes, bobcats, foxes, and sometimes bears) must present an additional amount of physical and psychological struggle wild animals have to deal with.

Relevant Considerations for the Animal Standpoint

Critics may object here and claim that animals do not have a standpoint for us to consider. However, and as the next section summarizes, an amazing as well as an expansive body of empirical evidence exists and firmly establishes that we are surrounded by other beings with their own minds and sentience, the ability to think, making decisions, and to experience emotions such as happiness, gratitude, grief, pain and suffering.

Our level of knowledge and awareness is such that the burden of proof is squarely on those who doubt what science, ethics and common sense have known for a long time—animals are subjects who have rich conscious, emotional and conative lives, experiences and abilities that are often on par with human abilities and sometimes exceeding them. Today, consensus in animal ethics and advances made in science all point to the fact that nonhuman animals are agents of their own world, "fellow travelers," as American writer Henry Beston[14] called them once.

Thus, the more relevant and interesting questions here, as posed by German and South-African humane educator Kai Horsthemk, is what exactly comprises the many perspectives of animals and "how can we speak with authority on the plurality of perspectives that make up the animal standpoint, also given the enormous diversity of nonhuman animals?" Horsthemke writes that as subjects, nonhuman animals have a particular viewpoint from which they perceive the world, and they have interests and because nonhuman animals are conscious and sentient subjects, we can ask what their interests are.[15] Thus, the important questions to ask here are: What are the interests of other animals independently from those of exploiters or humans generally? Would animals want to live or die?[16] In the context of the NAM, we need to ask: What are the interests of wild animals *independent* from those of sportsmen?

Represented by the NAM, wildlife management is the only social organization I am aware of that openly proclaims that killing individuals is not only good for them but that wild animals even benefit from lethal management. Consider the following comparison of an approach to human-animal relations from a human supremacy perspective versus one that considers the animal standpoint. "In a sense, the furbearers pay for their own management and benefit from it" is a statement made by Tom Krause,[17] a former president of the National Trappers Association. This is a statement from a human supremacy view—the opaque view of the dominator and oppressor culture—that is blind and ignorant to the interests and needs of those they subjugate and exploit. In total opposite, considering the perspective of the affected animals, would wild animals really choose management by humans? Indeed, would they choose lethal management as claimed by Krause? What are the interests of wild animals independent from those of sportsmen's—their exploiters and killers?

These are some of the morally significant aspects of the animal standpoint. And with a modicum of imagination and empathy, we can arrive at the conclusion that all animals, humans included, value survival and prefer to experience life without harm but instead with joy, happiness and contentment. This book points out the importance of empathy and for us to recognize and live accordingly to the shared needs of all sentient beings to live in peace, without being subjected to physical and psychological discomfort, stress, physical and psychological trauma and their lives prematurely ended. "These needs and preferences," states Horsthemke, "are essential features of the animal standpoint that are easily determinable and that lend themselves to being taught to children and adult learners, to engage with, to understand and to value."[18]

Horsthemke makes another, important point: Given that animals cannot tell us what they think of us, animal standpoint requires interpretation on our part and herein lies the challenge of discerning our own emotions from those likely experienced by animals facing situations of insurmountable odds. It continues to be my experience, likely shared, to feel sadness, anger and outrage when witnessing cruelty and crimes against animals—from chickens, cows and pigs brutalized on large-scale factory operations and in slaughterhouses, to an elk shot for trophy or animals caught in a trap or snare that slowly suffocates them. "*This*," writes Horsthemke, "the pain and the failure to make sense of a situation that is—more often than not—inescapable, desperate, is part of what constitutes the animal standpoint; and this is something we can understand, learn and teach about, on the basis of our imagination and empathy."[19]

The human supremacy worldview has imprinted itself in the collective consciousnesses of global humanity, and we have fallen victim to a paradigm of differentiation driven by the conviction that we have more rights to live and flourish on this planet than any other species. As our own population continues to explode, and we intensify and increase our exploitation of other animals, we have created a tipping point of an imminent mass extinction crises. Already in 1967, Lynn White Jr., a medieval historian, warned that an anthropocentric value system lies at the root of our ecological crisis, a view he largely blamed Christianity for promoting.

Major narratives about human society, nations and cultures exclude other life forms, and continue to define animals as creatures devoid of reason, autonomy, and inherent value, so they can be abused and used without mercy or compassion. In most disciplines, from anthropology to wildlife biology, conservation and management, animals are presented as passive objects determined by biology and genetics, devoid of agency, individuality, lacking subjectivity, morality and culture of their own. As Best has commented, the assumption here is that only humans are conscious, self-directing, and purposeful agents, and animals are reduced to being resources, commodities, interchangeable, the "'raw'" material" for human thought and action, and mere means to human ends.[20]

Curricula in schools do not teach humans to understand animals as sentient and autonomous beings, instead, 4-H school programs teach children to raise and care for an animal until a certain point, at which s/he/they must let go and send the young cow, sheep, pig or other animals on a horrific journey to slaughter—this is one of the biggest betrayal on all accounts, where the child begins to internalize the callous belief that the being so lovingly cared for, is "just a cow" and serves no other function than to end up as food. "Not to worry, it's just an animal" or "that's what they are here for" (e.g., addressing the imminent slaughter of this particular cow, pig or a deer as their only justification for their existence), we often hear from those who benefit from the exploitation of animals. This is an example of how we ask humans not to be animals, yet another aspect of a false perspective of how we exist in the world.

One of the consequences of our socially constructed dualism and the view of other animals as inferior is represented by the existence of the Animal Industrial Complex.[21] Species separation creates violence and speciesism strips animals of all intrinsic value to reduce them to instrumental value, to mere tools and objects whose cosmic purpose is to satisfy human interests. Every year on this planet, trillions of sentient chickens, cows, lambs, pigs, turkeys and other 'food' animals are confined under horrific conditions, grossly exploited, mutilated, genetically manipulated, pumped full with antibiotics and other drugs, tortured, and killed in the industrialized food system. Dr. Karen Davis, founder of United Poultry Concerns, when visiting an industrialized, large-scale chicken farm, stated that, "[T]hese birds are in a condition of pain, lameness, fear, bewilderment and learned helplessness. They are imprisoned in alien, dysfunctional bodies, in confinement facilities full of filth including atmospheric poisons and contaminants. What will it take for our species to realize that treating creatures this way is a crime, an iniquity?"[22]

Other animals are forced to endure miserable lives before they are killed for human interests such as clothing, entertainment and medical research and product testing.[23] Globally, wild animals are killed for their body parts (trophies) while others are kidnapped from their homes, thrown into cages, and transported to so-called "wet markets" where they await their brutal slaughter. In addition, tens of millions of semi-wild animals are kept under miserable conditions on so-called fur farms, large-scale operations that keep these animals—bobcats, foxes, mink, and others—for the sole purpose to kill them for their fur and to support the global fur trade.

The NAM has fully absorbed and continues to operate within a human supremacy paradigm. Correlated with the denial of subjectivity for animals is their placement as a cultural "Other," in opposition to humans. The NAM uses the strategy of othering as a powerful way of excluding animals from our moral concern. First, it presents the historical account of conservation as a sole human-centric enterprise ignoring the fact that we have always been in the presence of wild animals, and also omits the existence and experiences—or standpoints—of individual wild animals. Second, it omits the ways in which wild animals (albeit involuntarily) have contributed to the body of knowledge that wildlife 'professionals' have come to rely on. Also, sportsmen's past "victories" and "successes" have all been built on the reproductive work done by female animals, and the lives and deaths of hundreds of millions of individual wild animals—the game animals, predators, furbearers and other wild animals. As Best reminds us, "Just as the story of ruling classes cannot be understood apart from their relations to oppressed classes, so too human history cannot be grasped outside the context of the powerful determining effects of animals and nature on human society."[24] With respect to the NAM, this means that the ruling class of sportsmen's account of conservation's history is distorted and incomplete without at least acknowledging the existence and role of wild animals.

Similar to historic slave owners, whose wealth would not have been possible without the exploitation, forced labor and violent deaths of fellow human beings, the disciplines' existence, knowledge, and wildlife 'professionals' careers, and professional status and livelihoods, would have not been possible (at least not the way it has come about) without the annual death toll of hundreds of millions of wild animals whose lives have been and continue to be forcibly ended by sportsmen and agency personnel. These human 'accomplishments' are built on the backs of nonhuman animals, while at the same time, their existence and human impacts on these individuals, their mates, families and ecosystems are omitted, and their existence, death and roles remain entirely omitted.

Perhaps most obviously, we see the 'othering' of wild animals in the NAM's strategy of dividing the planets' diversity into humans and natural resources (opposites). Categorizing wild animals 'natural renewable resources' for sportsmen to use, "falsifies the psychological, intellectual, social and moral complexity of animals; it also fails to grasp how animals change environments, and to explain that they are not merely changed by them."[25]

Taking a very authoritative position in defining the type of view and treatment of wild animals, the NAM has declared that what matters in conservation are the populations and species but not the individual animal. In fact, the individual animal has been declared a "natural *renewable* resource" and irrelevant, an approach to conservation that leaves the individual wild animal vulnerable to human actions and invisible, yet it is the individual animal who is impacted by the NAM's practices including violence and death.

High Stakes and the NAM's Claims of Truth

The NAM's claims, especially because they are not supported by ethics and science, should be scrutinized carefully. One way is to look at the stakes certain human groups hold in making truth claims. In order to convey my point, I rely again on an analogy from the past as provided by Horsthemke: "Slave owners believed that slaves were thought of as naturally constituted for slavery as a matter of metaphysical fact,"[26] he writes, and further points out that this 'truth claim' was made solely on the basis of social interests. Similarly, the stakes for NAM promoters are extremely high when they are making claims about the irrelevancy of the individual animal, and the alleged need for hunting and trapping (killing) wild animals as necessary, and as management 'tools.'

Again, Horsthemke (2018), alerts us to the difference between *knowing* and ways of *claiming to know*, and states that what we are dealing with are "androcentric, Eurocentric and anthropocentric claims of knowledge, i.e., ways of *claiming to know*—which are indeed, governed by hegemonic interests and power relations .... It is knowledge claims, but not knowledge as such, that may result from social relations and that are often ideologically biased." And he further explains that "[K]nowledge claims may be mistaken, involve false beliefs, or be only minimally justified" ... and these, if they are genuine knowledge claims, including those made by experts and authorities, need to be checked against the facts or prove their mettle in the give-and-take of democratic deliberation and reasoned, scholarly argumentation."[27]

This is why I call the NAM's use of ethics and science "selective" because its promoters only use the two disciplines' knowledge that support their goal of exploitation and not the entirety of what is known about animals, and their claims have not been scrutinized sufficiently, if at all. I do believe that NAM's claims of knowledge about wild animals' status as objects and passive targets for sportsmen's actions will not survive such a fact check nor democratic deliberation.

Traditional wildlife management imposes an artificial and consequently deadly layer of regulations, rules, codes, laws and policies onto the world of unsuspecting and defenseless wild animals, all in service of meeting interests of sportsmen as actors, while treating animals as passive objects and the recipients of death sentences in the name of recreation. This endless cycle of sportsmen invading the world of wild animals on a regular basis and their obsessive need to kill sentient beings is truly questionable human behavior (to put it mildly), especially when the opposite usually does not happen, and wild animals try to stay alive and minding their own business while staying away from humans as best as they can.

The lives of wild animals are already rather challenging and often individuals are involved in life-or-death situations that make survival uncertain; they really do not need the added human-caused stressors, such as being chased by hunters/shooters and trappers and/or their hounds into death; they do not need the pollution, and noise that comes with these human intrusions and disruptions of their lives, nor do they need the confusion and grief that comes with getting injured and experiencing suffering and pain; they can do without the trauma and the entire mayhem caused by sportsmen causing someone to losing a mate, friends, parents and/or children.

Instead of supporting wild animals in their existing struggles to survive and cope with the multitude of human-caused threats, traditional wildlife management adds yet more sources of stress, trauma and mortality for wild animals (see Chap. 7). Due to the multitude of anthropogenic and natural threats and deadly risk factors that burden wild animals, especially posed under climate change and global destruction of biodiversity and natural habitat, we should no longer view hunting and trapping in isolation but rather as part of an overwhelming set of factors detrimentally impacting their lives and threatening their survival.

Understanding Animal Standpoint Theory

Horsthemke again provides further guidance on how and to what extent we can lay claim to epistemic authority in order to represent the world from the situated perspective of animals. The experience of other-than-human animals understood on the basis of evolutionary theory, ethology, and the like, provides a basis for potentially more complete and less distorted knowledge claims than do anthropocentric orientations. Nonhuman animals—domestic and wild—use unambiguous and complex forms of communication, intelligence, emotions and thought, and behave in purposeful and intentional ways. "They have obvious needs and preferences, like a desire for freedom. They will attempt to escape, avoid noxious stimuli, and have been known to attack their tormentors."[28] Doubting the idea that moral agency, self-determination, and autonomy are indeed matters of degree, and that, just like virtuous humans, other animals too are "motivated by tenderness, affection, prudence, deference, altruism, aggression and dominance" tells us more about us and our belief in a hierarchy of moral worth (topped by humans) than it provides a true representation of all life forms. And these reflections are all central in animal standpoint theory:

> The fact that animals too are conscious individuals, subjects of a life that matters to them, that can be better or worse *for them,* who can be benefitted and harmed, individuals with needs, beliefs, desires, preferences and expectations, best accounts for the initial intuitions behind animal standpoint theory. That there is something that it is like to be a particular animal, who experiences life and the world from a particular perspective, vantage- or standpoint from which life and the world are perceived and experienced, a perspective that is communicated in a variety of ways and that is informed by more or less complex needs, beliefs, preferences, desires and emotions also constitutes a suitable range of topics.[29]

The will to survive is probably the strongest emotion that exists in humans as well as other animals. Freedom is life's currency for survival, and all living beings will invest into the fight to achieve it at all costs. We know from our own, human-to-human relations that in cases of human rights violations and crimes against humanity, oppression causes resistance. An example of this in non-human animals is Molly B, a cow who escaped a slaughterhouse, jumped fences and ran as fast and far away as she could. Luckily, her efforts were successful and compassionate humans took her to safety—a farm animal sanctuary where she got to live out her life, among other rescued cows, pigs, sheep and turkeys, all well taken care of by kind humans.[30] Many of us are aware of situations where other animals if given the chance of escape, will just do that—flee and try to attempt a safe distance where survival is possible. Animals clearly show resistance to human domination through leeriness, fleeing, running away and distancing themselves from humans, and expressing their undeniable will to survive and resistance to being killed.[31] As Best notes, animal resistance to human domination is everywhere, and he points out that this type of resistance as an indicator of animal agency has been erased along with the awareness of the constant manifestation of their will, choices, and desires for freedom.[32]

Minding Animals: Consciousness

In his last opus, *The Formation of Vegetable Mould Through the Action of Worms, with Observations on Their Habits,* Charles Darwin presented two aspects of earthworms' inner life—cognition and subjective experience. His meticulous observations of earthworms and how they constructed and inhabited their burrows and handled leaves and other objects (thereby judging how best to pull these into their burrows) caused Darwin to conclude that earthworms have a degree of intelligence. Thus, Darwin found mind where most other scientists at his time, dismissed its existence in other-than-human animals and believed there was none.[33] For the next 135 years, mechanistic models of science dominated conservation and all other fields of inquiry and largely prevented us from building on Darwin's knowledge.

However, over the last few decades, discussions in science and the humanities have begun on the importance of abandoning the Cartesian view of other animals as soulless, non-thinking, and non-feeling beings. Evolutionary biologists, cognitive ethologists, social neurologists and field researchers all have provided copious evidence for nonhuman animals' capacities for consciousness, intellectual lives, physical suffering and emotions, and urged viewing other animals as subjects with moral sentiments similar to human beings. This concept is not new—Darwin provided fundamental support for the notion of continuity prevailing between nonhuman and human animals. From there, Darwin proposed that differences between the species are differences in degree rather than kind; they are not black and white but rather, shades of gray.[34]

Today, the convergence of ethical consensus and scientific understanding of nonhuman animals has led to the grounding of animal defense and animal rights. The concept of sentience—the capacity to experience emotions, sensations and feelings all related to consciousness—has been a crucial component of contemporary Western thought since the English utilitarian philosopher Jeremy Bentham explicitly stated that the question with other animals was not "Can they reason?' or "Can they talk," but 'Can they suffer?" This premise has grounded reflections defending the rights and interests of animals since philosophers Peter Singer and

Tom Regan founded the contemporary field of animal rights and liberation, and the subsequent development of the feminist concept of care put forward by Josephine Donovan, Carol Adams and Marti Kheel. These scholars provide philosophical support for the view of nonhuman animals possessing inherent value as sentient subjects of a life and deserving of having their major interests—well-being, autonomy, and life—considered and respected by humans. In this context, the work of American author and animal activist Jason Hribal further illuminates how animals resist their fate and helps us understand the autonomy of nonhuman animals.[35]

The fact that other animals share our capability to experience emotions and cognition has been well established since 2012, when a group of renowned scientists published the "Cambridge declaration of consciousness," which affirmed that all mammals, birds and other species have the neuronal architecture that underpins consciousness. Specifically, "nonhumans have the neuroanatomical, neurochemical, and neurophysiological substrates of conscious states along with the capacity to exhibit intentional behaviours."[36] The declaration essentially consolidated contemporary research findings, some of which corrected previously assumed hypotheses in science that leave the assumption of the uniqueness of consciousness as solely experienced by humans in the rearview mirror. By doing so, the declaration corrected commonly held beliefs about the hierarchical thinking of evolution that views animals as merely being less-than-humans and noted instead that the behavior, neurophysiology and neuroanatomy of birds and mammals are a striking case for parallel evolution of consciousness. In sum, the declaration, along with other research, not only moves animal experiences out of the shadow of human uniqueness and position them on par with *Homo sapiens* but opens the door to considerations that other animals may by far surpass whatever humans generally possess. For example, other species have sensory-perceptual capacities—such as olfaction in dogs, audition in cetaceans, or use of ultrasound by bats—that are far more sophisticated than what humans in general possess. Case in point, Lori Marino, a leading neuroscientist and expert in animal behavior and intelligence has stated, that "[C]learly, echolocation—the use of high-frequency sound for processing information—by cetaceans is more than a fancy receptor device [usually this ability is described in non-cognitive terms]. It is a highly sophisticated cognitive capacity that humans lack—and all that that implies."[37] This raises significant concerns about to which degree our human supremacy prevents us from recognizing and appreciating the abilities of other animals, especially when this would finally invalidate our inflated sense of exceptionalism?

Neuroscientist Antonio Damasio sees animals as having a core consciousness that enables a sense of self, independent of reason and language.[38] Neurons, the brain's building blocks, which were long thought to be exclusive to humans, are similar across animal species; unique spindle neurons, which play a leading role in social and emotional intelligence, vocalization control, and even facial expressions, have been found in the cerebral cortex of humpback whales, while mirror neurons, which make it possible for humans to recognize the actions and intentions of others, have been found in other species as well. These findings are not limited to birds and mammals, and research suggests that some insects possess ganglion structures that may enable the capacity for subjective experiences—pollinators such as bees, for example, are not only able to learn certain behavior but also possess the ability to pass it on to their conspecifics.[39]

A 2021 review[40] by animal welfare experts in the United Kingdom (UK) confirms lobster, octopi, squids, crabs and shrimps are sentient sea creatures with a central nervous system, who can and do experience pain. The reviews' findings correct previously held beliefs about nonhuman beings as inferior—this time the assumption that only animals with backbones are sentient. As part of the implications of the investigators' research, all decapod crustaceans will now be protected under the Animal Welfare Bill in the UK.

Marc Bekoff, professor emeritus of Ecology and Evolutionary Biology, and one of the key scientists advancing compassionate conservation as a new approach to our relations with wild animals, has written extensively for several decades about the needs and interests and the lives of other animals. Pointing out the importance of non-invasive methods, he has concluded that the more we study animals the more we learn about their rich emotional lives and cognitive skills making up their amazing world. Indeed, chickens are courageous and empathetic to others' distress, so are mice and rats; foxes use the Earth's magnetic field to measure distance and direction when they pursue prey under the snow; new Caledonian crows make sophisticated tools, more so than chimpanzees; wolves, coyotes, and foxes, all seem to have species-specific moral codes that guide their intra-species behavior, such as the concept of fairness and altruism, the latter also being commonly observed in insects such as bees, wasps, and ants (e.g., see Nurse and Wyatt (2021, pp. 75–80) for examples of several studies showing altruistic and pro-social behavior inherent in nonhuman species).

Bekoff and his colleague Jessica Pierce, a philosopher and teacher, argue that cooperation is seen in a wide variety of animals, including rooks, hyenas, and wolves. And there is evidence that some animals help animals of different species.[41] For example, biologist Bernd Heinrich found that ravens lead wolves to elk carcasses that ravens cannot process on their own, thus the ravens let wolves tear the carcasses apart before they start eating themselves.[42] Trail cameras have captured coyotes and badgers having fun traveling together to hunt mice and voles as partners.

Along with Bekoff, other ethologists, including Carl Safina (2015),[43] and Frans de Waal have provided conclusive evidence in support of nonhumans' rich emotional and cognitive lives and how they think, solve problems, understand concepts, and communicate, and have helped us to acknowledge that the lives of nonhuman animals are more complex than ever understood before. Primatologist and animal rights advocate Jane Goodall as well as Cynthia Moss, who have observed primates and elephants respectively for decades broke through the ceiling of viewing wild animals just as numbers, and instead, knew these animals by their personality and hence named them individually and by family membership, now a common practice in field research.

Bekoff's research finds that emotions have evolved in numerous species as an adaptation, a social glue to bond nonhumans, as catalysts and regulators of social encounters and as a measure of protection. It further shows how different species of animals, including coyotes, elephants, and macaques, possess complex social behaviors that include a sense of justice, cooperation, empathy and a readiness to aid their companions in distress. Baboons engage in deliberate acts of deception, magpies exhibit grief, and dolphins will band together to keep an injured fellow afloat.

Bekoff concludes that morality and empathy are evolved traits shared by humans and nonhumans (see Chap. 8). Similarly, Dutch primatologist and ethologist Frans de Waal, has demonstrated primates' facility for conflict resolution, cooperation, fairness, food sharing, an awareness of the permanence of death, and the capacity for empathy.[44]

Wild animals engage and are embedded in a plethora of complex social bonds. For example, wildlife biologist Doug Chadwick, while tracking wolverines (*Gulo gulo*) in Montana's Glacier National Park, discovered two sets of paw tracks in the snow that indicated that two wolverines had been trotting side by side. They turned out to be a father and his one-year-old son who had already left his mother, both "hanging out together," perhaps spending quality time together, something Chadwick called "a wolverine version of joint custody." Known to be mountaineers extraordinaire, Chadwick's observation corrected previously held beliefs that assumed wolverines were strictly solitary except during mating seasons and considered to be "too volatile and vicious to tolerate company of any kind." Instead, and thanks to Chadwick, we now know that wolverines actually maintain long-term, social bonds with one another.[45]

Some scientists are exploring the question, "How do animals think about death?" Opossums are known to "play dead" when they realize that a predator is nearby, and a growing body of science shows that animals have a concept of death and that they mourn loss and also grieve.[46] Safina, author of *Beyond Words: What Animals Think and Feel* writes that predators, among many other animals, "must understand the difference between alive and dead when they kill their food." Safina adds, "humans have many concepts of mortality—karmic wheels, eternal life and so on—most of which conflict, which indicates most of them are wrong."[47]

In 2016, a group of scientists reported on 16 observed cases of free-ranging marine and terrestrial mammals mourning their dead, further corroborating that adults mourning their dead young is a common and globally widespread behavior. Their findings ranged from African elephants, who are known to stay in an evident state of distress near a dead or dying calf, to primates carrying dead infants, to wolves burying dead pups.[48]

In 2018, the plight of a killer whale held the online human world in its grip, where clearly stricken by grief over the loss of her baby, the killer whale mother was observed carrying her dead calf for at least 16 days before finally letting go.[49]

Rick McIntyre, a longtime wolf observer and an expert in wolf behavior, relays the story of a wolf in Denali National Park in Canada who had been killed by a sportsman, her mate returning to the kill site for a long time until the day he was also killed by a sportsman (personal communication). In his book, *The Reign of 21,* McIntyre tells the story of Yellowstone National Park's famous male wolf 21, who had spent most of his life with his mate (female 42) with whom he had raised several 'litters.' After the disappearance of his mate, wolf 21 searched for her for a long time only to decide to return to the site they had spent most of their time together as a couple, where he died.[50]

We are now aware of the occurrence of Psychological Trauma Stress Disorder (PTSD) in many wild animals[51] notably in free African elephants, who have witnessed the massacre of their family members (so-called 'culling"). Behavioral observations by Wildlife biologists Jay Mallone and Paul Joslin reported PTSD in Tenino, a wild wolf, who was placed in captivity after being traumatized by USDA actions.[52]

Others, who observe or work more closely with nonhuman animals, for example veterinarians, wolf observers such as Rick McIntyre in Yellowstone National Park, animal rescue and animal and wildlife sanctuary staff, wild animal rehabilitators, and those of us who share their lives with multiple dogs, cats, horses and other animal family members can usually easily recognize individual nonhumans' feelings, personalities and moods. Videos following the liberation of chimpanzees from longtime captivity and exploitation in research laboratories, show these animals hugging each other the first time they step outside into the welcoming sanctuary landscape; a 2021 video on the spectacular rescue operation of 100 moon bears, liberated from so-called 'bile farms' in China, show the bears' individual reactions from initial hesitation and fear to being overjoyed when they can move freely and feel grass under their feet for the first time in their lives.[53]

Finally, even the most cursory look on the internet shows an endless number of videos capturing moments in animals' lives such as bears sliding down a steep snowy hill just to climb up and do it again, anecdotes from animal activists, observations by field researchers, longtime wildlife watchers, animal sanctuary staff and scientists' field notes, all show almost on a daily basis what other animals share with us. In addition, as Max Foran reminds us, there is always the unknown, for example, "what impelled a group of seventeen wild turkeys on a Massachusetts' street to circle a dead cat in almost ritualistic fashion?"[54] An interesting question to ask ourselves when considering the lives and behaviors of other animals is, what actually constitutes our right or need to know?

The consciousness debate emerges as pivotal because it is linked to moral worth or ethical treatment, both of which are denied to other animals on the grounds that animals lack those mental abilities and traits that make us humans. Others dismiss animal consciousness and its moral implications by claiming that it can never match that of humans. One problem with this is that animals continue to be measured in human terms. "Why should humans have a monopoly of a crucial quality such as consciousness?" asks Horsthemke, and points to the importance of consciousness and the ability to perceive pain for the evolution and survival of animals.[55] If we have learned from our intra-human relations, it is this: Being different does not mean inferior or less.

The relevance and importance of this research is that it shows that there is no moral gap between humans and nonhumans. Therefore, the idea that human interests are above that of other animals is no longer defensible from an ethical or a scientific perspective. Yet, the NAM along with many if not most contemporary societies and governments on Earth define humanity as separate from (and in certain ways in opposition to) the rest of nature. We continue to live under the hegemony of human supremacy and speciesism. A leading scholar and critique of anthropocentrism, Eileen Crist, defines human supremacy as "the widely diffused belief-system that human beings are a superior life-form that is authorized to overpower and do what we will to all other life forms, to all places, and (in our time) to Earth as a whole." Crist continues, "Most people do not espouse to this belief system consciously; they are corralled to live by its precepts and trained to assume they are sound."[56]

Changing Public Values: Toward Mutualism and Animal Rights

Concomitant to advances in sciences and ethical consensus in animal liberation and ethics and critical to this book, are changing public values with respect to the view and treatment of other animals. A 2018 study showed that U.S. citizens increasingly humanize wild animals and supported the idea that wild animals are no longer seen as a commodity and that they should have similar rights as humans. Already in 2015, a Gallup poll showed that "Almost a third of Americans, 32%, believe animals should be given the same rights as people, while 62% say they deserve some protection but can still be used for the benefit of humans. The strong animal rights view is up from 2008 when 25% thought animals' rights should be on par with humans.'

An ongoing value shift in the U.S. has led to growing public support for the preservation of nonhuman nature, including larger landscapes, especially of western public lands. The value shift toward mutualism reflects an egalitarian ideology of inclusion and "the view of wildlife as capable of relationships of trust with humans and defined by a desire for companionship with wildlife."[57] And furthermore, individuals with a mutualist orientation view wildlife as an extended family deserving care and rights, and, as such, are "less likely to support actions resulting in death or harm to wildlife, more likely to engage in welfare enhancing behaviors for individual wildlife, and more likely to view wildlife in human terms."[58] Interestingly, research has long shown a link between human-wildlife relationships and human-human relationships within societies. Indeed, findings have confirmed that with the transition from hunter-gather societies to agriculture-based societies came "a shift from trust and mutualism to domination in human-human relationships *and* in human-wildlife relationships." Based on their studies, social research scientists, Michael Manfredo and Tara Teel propose the return to trust and mutualistic views of wildlife in post-industrial societies, finding such a shift likely given the changes that have already been occurring. Some examples include a decline in concern for material well-being, an increase in egalitarian relations, and an increased desire for belonging and self-esteem. The authors conclude, "These trends in human-human relationships set the stage for a re-emergence of trust views toward wildlife and a decline in views of dominance and mastery."[59]

The ongoing value shift among the public involves emotional attachment and a sense of empathy for wild animals, qualities that indicate a growing disconnect in values between the public and the wildlife profession, which still holds on the human supremacy view of animals as "natural renewable resources" to be used by humans. The public's changing valuation also means a growing potential for increased social conflicts on wildlife issues, characterized by clashing values and a strong desire to intervene in policy through mechanisms such as ballot initiatives to ban trapping. This leads to a hardening of fronts on both sides with no solution in sight.

In addition to the historic and ongoing public repugnance of trapping, societal consensus has never been stronger and more aligned than at present, calling into question the justification of killing wild animals for fun, trophies, recreation and commerce. Trapping in particular, has a long history of public repugnance and opposition, and survey after survey has shown that the public does not support trapping, especially when educated about non-lethal methods. For example, a nationwide poll on trapping in the National Wildlife Refuge System, commissioned by the Animal Protection Institute (API) confirmed previous findings that showed the public is opposed to trapping because it inflicts pain and suffering to the animals caught, and that a majority (58%) of the 800 Americans interviewed, would support a ban on all trapping.[60] A survey of 3127 conservation professionals, including members of The Wildlife Society asked whether leghold traps should be banned. 46% of them indicated yes, 39% no, and 15% had no opinion. These professionals cited pain and stress and harm to non-target species as the two primary reasons, followed by the view that the leghold trap is not a necessary tool for wildlife management.[61]

NAM proponents protect their interest in the exploitation of wild animals in opposition to the public's changing relationship with wild animals, which is moving progressively toward life-confirming values. While NAM promoters view the growing interest of the public in liberating animals from the shackles of human exploitation, as a threat to their hunting [and trapping] heritage, it is clear that radical improvements for animals will not come from the inside of the dominator culture. Commitments to this human supremacy ontology make the NAM and the discipline of wildlife management increasingly look like a monolith of death amidst a sea of progressive, life-affirming changes among a general public increasingly turning toward an emphasis on greater care and compassion for wild animals.

The NAM and Oppression: Wild Animals Need Representation

As the reader will learn in Chap. 3, the history of the NAM has strong roots in elitism, colonialism, patriarchy, speciesism and human supremacism. These are different but interlinked manifestations of oppression that range from the individual to a collective level. In her book, *Powerarchy: Understanding the Psychology of Oppression for Social Transformation,* psychologist and social justice advocate, Melany Joy, has termed this "powerarchy," the overarching belief system that informs all oppressive systems. This, she describes as "a nonrelational system that is organized around the belief in a hierarchy of moral worth—the assumption that some individuals or groups are more worthy of moral consideration than others—and that is structured to maintain unjust power imbalances." Joy further explains that manifestations of oppression which include rape, child abuse, war, factory farming, racism, and speciesism—all spawn from this "belief in a hierarchy of moral worth" as the mindset that drives all oppression. Characterized by a denial of dignity and the violation of one's integrity, Joy calls the belief in a hierarchy of moral worth deeply "dysfunctional."

Hence the ideology of speciesism, when philosophically deconstructed, is just like other forms of supremacy and proves to be morally indefensible. And finally, Joy identifies the belief in a hierarchy of moral worth as one reason why we do not recognize certain individuals as *victims* of oppression, even when they are. Humans and nonhuman animals need to hold a certain status in human society in order to be recognized as vulnerable to victimization. Joy provides the example of girls and women, who had to experience sexual assault and PTSD not recognized until the feminist liberation movement of the 1970s elevated the social (and moral) status of females. In the same vein, the status of nonhuman animals—domestic and wild—must urgently be elevated so that their oppression and their rights become visible and recognized by human society.

The NAM is a major part of society's structure of oppression and the animal standpoint illuminates the NAM as a system with power relations that keeps wild animals subjugated and disenfranchised, while allocating all power over wild animals' lives to humans/sportsmen. Yet we know that, as subjects, other animals matter to themselves, to their friends, mates and families, and they should matter to us as well. Animals are conscious and sentient beings with interests, yet they have no consistent representation of their rights to exist and survive, to stay alive and be unharmed by humans. Wild animals are part of our society, or we are part of their nature. Either way, as victims of systemic violence, animal representation, the essence of the animal rights movement, should be included in social justice movements, not on the periphery of moral concern but front and center. Implementing this vision, however, will require us to change our belief system and our thinking and feeling about other sentient beings with whom we share the planet.

Consensus in ethics and advance in science both support animal rights. They do not support the NAM's inherent assumptions grounded in human supremacy. The rest of this book deconstructs the North American Model of Wildlife Conservation in order for us to envision changes both in perspective and policy.

Notes

1. Donovan, "*Caring,*" 361.
2. Wood, "Feminist."
3. Best, *Politics,* 3.
4. The Frankfurt School is a far reaching sociological and philosophical critical theory of modernity and society.
5. The Institute for Critical Animal Studies (ICAS).

6. Best, *Total Liberation*, 1.
7. Deckha, "CAS and Animal Law," 216.
8. Ibid.
9. Slicer, "More Joy," 23–24.
10. Foran, *Subjugation,* 27.
11. Paquet and Darimont, "Wildlife, Conservation," 178.
12. Almiron and Fernández, "Including."
13. Horta, "Concern," 85.
14. Beston *Outermost House.*
15. Horsthemke, *Animal Rights Education,* 16.
16. Best, *Politics,* 16.
17. Krause, *NTA Handboook.*
18. Horsthemke, *Animal Rights Education,* 211.
19. Ibid., 215.
20. Best (2014, p. 5–6).
21. Harmonist, "Animal Industrial Complex."
22. Davis, K, Open Letter."
23. Food and Agricultural Organization of the UN, FAOSTAT.
24. Best, *Politics*, 4.
25. Ibid., 5.
26. Horsthemke, *Animal Rights Education.*
27. Ibid, 209.
28. Ibid.
29. Ibid., 216.
30. Kuglin, "Famous Cow."

    I previously relayed this story in my chapter, "Earth's Animal Nations: It's Time for Human Compassion," In: *Visions for a Post-Covid World: Defining a Radically New Normal*, edited by Joe Gray, Dixi Books, 2021, Palmers Green, London, England, pp. 15–28.
31. Best, *Politics.*
32. Ibid.
33. Crist, "Inner Life."
34. Bekoff, "Animal Minds."
35. Hribal, *Fear.*
36. Low Philip, "Cambridge Declaration."
37. Marino, "Trans-Species."
38. Damasio and Meyer, "Looking Glass," 167–8.
39. Foran, *Subjugation.*

40. "British Study," NPR.
41. Bekoff and Pierce, *Wild Justice.*
42. Heinrich, *Mind of the Raven.*
43. Safina, *Beyond.*
44. de Waal, *Age of Empathy.*
45. Chadwick, *Wolverine Way*, 26–27.
46. Bekoff, "Dogs, Coyotes and Opossums."
47. Yeoman, "Animals Grieve."
48. Reggente et al., "Nurturant Behavior."
49. "Killer Whale," BBC News.
50. MycIntyre, *Reign of Wolf 21.*
    Note: See this website with a photo of wolf 21 and his true love, wolf 42: Living on Earth: The Reign of Wolf 21 (loe.org).
51. Levy, "Animal and PTSD."
52. Bekoff, *Animal Manifesto,* 15.
53. AnimalsAsia, "101 Bears Home."
54. Foran, *Subjugation*, 18.
55. Horsthemke, *Animal Rights Education.*
56. Crist, "Abundant Earth" (Interview by Marc Bekoff).
57. Teel and Manfredo, "Understanding Diversity."
58. Ibid, 130.
59. Ibid, 194–195.
60. Born Free USA, 1999.
61. Muth et al., "Unnecessary Source of Pain."

3

The North American Model of Wildlife Conservation

Settling the West and the Era of Unregulated Wild Animal Slaughter

To understand the current system of government-run wildlife management in the United States, we need to look at the history of the human value system with respect to wildlife in North America over the past 150 years, an era that encompasses the greatest slaughter wild animals have ever experienced. White settlers from Europe brought with them a decisive human supremacy worldview that categorized wild animals according to their perceived usefulness—some animals were 'food,' others were to be hunted for fun or trapped to procure fur, and predators such as wolves were the enemy and greatly despised as 'vermin' to be destroyed and eradicated by any means.

With no ethical consciousness toward wild animals other than a 'use' mentality, settlers—in particular, market hunters and fur trappers—engaged in the greatest wildlife massacre the continent had ever seen. Wild animals were defenseless against this human assault. After technological innovations such as the manufacturing of steel traps, repeating rifles, and the transcontinental railroad line (the latter opened in 1869)

A. Heister, *Beyond the North American Model of Wildlife Conservation*, The Palgrave Macmillan Animal Ethics Series, https://doi.org/10.1007/978-3-031-14149-2_3

leaving few areas inaccessible to commercial hunters, animals sought for their fur, plumage, or meat had nowhere to hide either. In the absence of any effective regulations to control such a free-for-all, staggering numbers of wild animals were killed in the course of just a few decades.

In part motivated to eliminate the cultural and subsistence needs of native peoples, settlers liquidated the American bison (*Bison bison*)—reducing their numbers from an estimated 30–60 million to approximately one thousand in all of North America in 1890. Enormous numbers of elk, deer, pronghorn, and bighorn sheep were also killed. Entire bird species were gunned down to meet the demands of the millinery trade— including the passenger pigeons, whose populations dropped from 3–5 billion to zero in just a few decades and whose last member, named Martha, died in captivity in 1914. Commercial trappers killed tens of millions of beavers in order to supply enough animal pelts to satisfy the Euro-centric dictates of fashion and human vanity. While the beaver populations have been estimated at 60 million before the arrival of European settlers, by 1900, commercial trappers had exterminated beavers in much of their U.S. range and their total numbers in the United States were reduced to fewer than 100,000 animals. Heroic restoration efforts during the 1900s included the live capture of beavers to transport them across states and even parachute them from aircraft into remote regions where they could repopulate without being hunted or trapped. Restoration efforts were financially supported through the Federal Aid in Wildlife Restoration Act, passed in 1937. However, by the mid-1950s, the trapping of beavers resumed.[1]

Roosevelt and the Rise of Sport Hunting

Today, Theodore Roosevelt (president of the United States from 1901–1909) is known as the president who saved more than 230 million acres of public lands including national parks, wetlands, and wildlife refuges. The other part of his legacy is tied to his passion for sport hunting. A committed trophy hunter, Roosevelt once traveled even to Africa with his son and shot over 11,400 wild animals—including elephants and lions—as trophies and as 'specimens' collected for the Smithsonian Institute.[2]

Toward the late 19th century, a group of influential and wealthy sportsmen led by Roosevelt started to organize against market hunting, and in 1897 founded the Boone and Crockett Club (BCC), a trophy hunting organization in Missoula, Montana. Based on the belief that people need and will always use natural resources, including wildlife, their idea was to create a new system called 'conservation' for use of wildlife that would be regulated and guided by science. Facing the potential threat by society demanding complete protection of wildlife and thus an end to hunting, they created and then heavily promoted the concept of 'fair chase.' This maneuver helped sportsmen avoid complete protection of wild animals and a potential loss of hunting. At the same time, it elevated the new class of 'sport' hunters, who claimed to be hunting for personal reasons (never for profit) and followed a code of honor, which intended to assign them a 'distinguished,' social role within society. This occurred at a time of significant social upheaval, with the ongoing discussion of rights for the formerly enslaved, and the women's suffrage movement asserting that women had just as much right to inform the government through their vote as men did. With these changes threatening the traditional concept of 'manhood,' Roosevelt believed that hunting had a civilizing effect on men, and he advocated hunting as an outlet for a man's "'virile impulses" and thus helped to create a new image of masculinity that promoted the attribute of self-restraint.[3]

Managing Game/Wildlife for 'Sportsmen' Becomes Government Business

The Boone and Crocket Club's mission has been "to promote the conservation and management of wildlife, especially big game and its habitat, to preserve and encourage hunting and to maintain the highest ethical standards of fair chase and sportsmanship in North America."[4] At the same time, the highly influential club promoted laws to protect the freedom of every citizen to hunt and fish and established that wildlife should be owned by the public and managed in trust by government agencies.[5] Early government-based wildlife conservation in the U.S. was, therefore, established by a group of recreational hunters *for* hunters in response to the decline of game species caused by market hunters— and also as a means of maintaining an ability to hunt wildlife as the western part of the continent was colonized. The now widely accepted concept of "fair chase" arose from Boone and Crockett's successful efforts to promote hunting as a socially acceptable practice and promoting recreational hunting to counter the effects of market hunting and unrestricted subsistence hunting. From recreational hunting soon came the rise of 'sport' or 'trophy' hunting. "To facilitate such hunting ethics the BCC promoted restrictions on how many animals could be killed, the season of hunting and other changes that once implemented did result in a recovery of so-called game species like elk and deer." Sportsmen also structured varied forms of boards or advisory councils or fish and game commissions for them to have a voice in preventing a return of the devastating market hunting.[6] These agencies were (and still are) largely hostile to predators such as wolves and coyotes.

To Native Americans, wolves were essential partners, spiritually and literally. Without wolves, they believed they could not survive. But President Theodore Roosevelt saw the wolf as a scourge and declared that wolves should be eliminated. His proclamation led to a nationwide massacre, a cruel holocaust of killing by gun, trap, poison, even by burning wolves alive. Before the arrival of Europeans, wolves were estimated between 250,000 and 2 million. By the mid-20th century, a ragged remnant of a few hundred was left. As Roosevelt "fought aggressively against preservationist or non-use proposals"[7] he essentially used his influence to put a decisive end to preservation (protection nature, including wild animals from use), the second dominant strand of philosophy at the time, and one that John Muir (the founder of the Sierra Club) had fought hard for. Finally, Roosevelt and the BCC initiated and lobbied for laws that stipulated the 'user-pay" system we know today as the Pittman-Robertson Act, and the Federal Duck Stamp program. Thus, by replacing market hunting with the concept of sport hunting, historic sportsmen succeeded in *conserving* wild animals for human use, and at the same time *preserving* methods to exploit them. Sport hunters achieved putting state wildlife agencies in charge to manage wildlife, in particular game species, on their behalf. Today's government wildlife-management system developed under a worldview of strong human entitlement (the human supremacy worldview) that reduced sentient wild animals to a status of natural (renewable) resources serving human interests. The sportsmen's 'consumptive use' mentality towards wild animals, along with their claims of possessing the right to hunt and trap them for 'personal use,' is what is at the core of the NAM, and what its promoters seek to protect.

The North American Model of Wildlife Conservation: Narrative and Elements

The NAM consists of a set of seven elements that are variously described as "the seven sisters of conservation" or "tenets" or "principles," (hereinafter, referred to as "elements"). These are (1) Wildlife are a public trust, (2) Markets for game/wildlife are eliminated, (3) Allocation of wildlife is by law, (4) Wildlife can be killed only for legitimate purpose, (5) Wildlife is considered an international resources, (6) Science is the proper tool to discharge wildlife policy, and (7) Democracy of hunting is standard. They reflect selectively chosen historic events, ideas and actions taken—spanning the era between 1870 and the 1930s—largely by influential and wealthy sport hunters in response to the decline of wildlife/game species caused by market hunters. NAM promoters claim that precisely this crisis set in motion the inexorable rise of the concept of sport hunting and the associated new ethics of the American conservation movement.[8] In contrast, historic interpretations by publications outside literature that support NAM practices, find that this change in attitude was prompted less by a concern for nature than a desire to protect certain game species favored by recreational and sport hunters.[9] The BCC's mission statement (see above) attests to the fact that the issue was the safeguarding of recreational hunting, which of course required an end of market hunting and the survival of game species.

Collectively, the set of elements compose a narrative that assigns nearly exclusive leadership to sport hunters in the development of the entire conservation movement in North America[10] Today, leading advocates describe the North American Model of Wildlife conservation as "a system of sustainable development of a natural renewable resource that is without parallel in the world."[11] In an equally admiring tone, others state that the tenets have led to an "… evolved and shared system of conservation laws, principles, institutions, and policies that has enabled the successes of Canada and the United States in the recovery, management, and protection of wildlife and brought them global recognition."[12]

The three dominant threads—representing quite spurious claims—of the NAM's narrative are, that 1) sport hunters effectively ended the historic massive wildlife slaughter caused by market hunters, and are responsible for the restoration and continued conservation of wildlife; 2) hunting embodies the three pillars of wildlife management—habitat, wildlife populations, and people—and as such, it is not only an essential tool for wildlife management, but "the original lever behind the North American conservation movement" and 3) wildlife conservation is only possible with the funding, involvement, and commitment of hunters.[13] These are also the messages that are relentlessly communicated to the general public.

On the other hand, the role of trapping in the NAM's narrative is notably absent or downplayed. We hear emotional statements by promoters such as "It's the man with the gun, and the man with the rod that has substantially supported and rescued wildlife in North America."[14] These rescue efforts apparently do not include the man with the trap. Yet today recreational and commercial trapping is clearly subsumed under the NAM's blanket, despite rather obvious contradictions between the latter and some of the NAM's core values. For example, the NAM vehemently opposes "hunting for profit" (market hunting) and claims to have eliminated markets for dead wildlife but sanctions the existence of commercial markets for furbearers (large-scale commercial 'fur farms,' along with the global fur trade). Another contradiction surrounds the definition of the *sportsman,* for which the concept of 'fair chase' is crucial. Yet, there is no fair chase in trapping animals, as the trapper neither 'knows his target,' nor is the trapper even present at the time an animal gets caught in a trap or snare. The number of unintended targets injured or killed in trapping practices is concerningly high. Trapping devices are often laced with either food as bait (often body parts of a previously trapped animal) or treated with lure (often urine or gland excretion from animals held in captivity on fur farms).

In their most recent, 2020 issue statement, The Wildlife Society (TWS) has taken a critical, yet contradictory viewpoint of baiting wild animals. TWS defines 'baiting' as "the act of intentionally placing food or nutrient substances to manipulate the behavior of wild species for the purpose of, a) attracting wildlife to a specific location to enhance hunter harvest, trapping, or viewing opportunities," among other purposes listed. TWS warns that "public baiting and feeding of game is often detrimental to wildlife resources" while at the same time, seemingly approving of baiting practices when "small-scale baiting is thought by some to be critical to hunter and trapper success as well as achieving harvest goals of some wildlife management agencies."[15] The idea of baiting animals does not seem to square with the ideals of fair chase, yet bait and lure are commonly used in trapping.

Furthermore, the two requirements of the sportsmen's ethics code—a quick kill to prevent unnecessary suffering, and that there should be no waste of the killed animal—do not apply to recreational and commercial trapping either, as many trapped animals suffer greatly before they die, and most trappers do not consume or otherwise use the animals' carcasses. A review of trappers' online forums reveals that animal carcasses are an unavoidable reality of their activity, and that these carcasses get mostly either buried in the woods out of the public's eyes, used as 'bait stations' to either shoot or trap more animals over, or—for the city-dwelling trapper—get dumped in counties' landfills. This is clearly the 'waste' of animals' bodies the NAM is allegedly so opposed to. As the online discussion board on "carcass disposal" shows, trappers' main concern about getting rid of the animal bodies is to discard them in places where the public doesn't see them.[16]

The original idea that these principles could be described as a model is generally credited to Valerius Geist, a Canadian wildlife biologist/zoologist of European descent. Geist articulated the NAM in 2001, together with Shane Mahoney and John Organ, all of whom are highly influential wildlife 'professionals' and among the most outspoken advocates for the NAM. Mahoney is considered an expert on the NAM, consults for trophy hunting organizations, and works closely with John Jackson III, ex-president of the Safari Club International, the world's most profitable and aggressive trophy hunting organization (see The NAM's Defense

Network, below). An extremely influential strategist for the sportsmen's movement, Mahoney once proclaimed the need for hunters to recruit more women because they "might possibly be the force which will softly murder the myth that hunting is evil."[17] After this call went out, a focused and strategic recruitment ensued, targeting women and thus increasing the number of female hunters from 13.3% of all hunters in 2007 to 20% (3,675,000 women) in 2016.[18] Mahoney's statement goes to the heart of the current public controversy surrounding traditional wildlife management: the strong conviction by a growing sector of the public that killing wild animals for recreation, trophies, and commerce, is morally no longer (if it ever was) acceptable.

The Seven Elements[19]

1. Wildlife Resources are a Public Trust

Referred to as the foundation of the NAM, this element reflects the understanding that wildlife, as a natural resource, is owned by the public and held in trust by the government for the benefit of present and future human generations. A common law principle, the "Public Trust Doctrine" applies to all natural resources and is the legal foundation for federal, provincial, and state agencies. Its historical development can be traced back to a Supreme Court decision in 1842 (*Martin v. Waddell*), which declared that certain resources—in this case oysters gathered by an individual on New Jersey mudflats—could not be owned privately. According to NAM supporters, a subsequent Supreme Court decision in 1896 addressing "illegal transport of hunted ducks across a state border firmly made wildlife a trust resource (Geer v. Connecticut)." Judge and legal authority Joseph Sax has traced the roots of the Public Trust Doctrine to Roman law, finding that Romans viewed wildlife as part of common property, which could not be privately owned and was for use by everyone. As suggested by legal scholar Susan Morath Horner, the category of things not to be owned by anyone (*res nullius*) were likely abundant and not appropriate for private possession and sale.[20] Though viewed as "wild and free," animals were always categorized as property; thus this tenet reflects the ancient principle that common law dislikes "ownerless" things. The English, who adopted Roman law after the Magna Carta (A.D. 1215), placed ownership of public resources with the king as trustee of certain properties for someone else.[21] In the United States, the 1842 *Martin v. Waddell* decision ascribed trustee status to the states. Sax has identified four fundamental concepts of public trust: common law, state law, property law, and a public right.

2. Markets for Game/Wildlife are Eliminated

Seen as a milestone and one of the most important policies in wildlife conservation, this element establishes that commercial use of wildlife, their parts or products should be prohibited based on the drastic decline of game populations caused by historic market hunting and the subsequent transition to recreational hunting beginning in the late 1800s.[22] Curiously, trapping for commerce (selling animal pelts to the markets) is exempted according to this tenet. Mahoney and Jackson (2013) share the view that the impact of trapping, which is seen as "labor and skill-intensive," did not amount to the same level as mass killing through market hunting, a belief that caused "hunters and anglers to feel that they had the most to lose." After all, as pointed out by these authors, the focus of this principle always was "to eliminate incentives that elicited *excessive* taking of wildlife, not to eliminate takings per se."[23] An increasing concern among NAM promoters is "a value system endorsing an animal-rights doctrine," which is seen as "a threat to the premise of public ownership of wildlife."[24]

3. Allocation of Wildlife is by Law

Allocating access to wildlife through hunting, fishing, trapping or otherwise, equally among all U.S. citizens is what Mahoney and Jackson have described as the "mortar that binds the entire North American Model."[25] This tenet again refers to the Public Trust Doctrine by emphasizing the democratic access to wildlife where 'surplus' (see the explanation of this misleading term in Chap. 6 wildlife is [sic] allocated to the public for consumption by law, and not by nobility, wealth, landownership, or other markers of elitism, as is the case in Europe. NAM promoters claim that it was legislation ensuring that every citizen could participate in the annual 'harvest' of wildlife not only provided the impetus for a developing conservation ethic but furthermore, egalitarian allocation also facilitated egalitarian cost-sharing through the "user-pays" principle. Furthermore, Geist points out, that by footing most of the bill for wildlife conservation, "hunters provide a benefit to society at large—the maintenance of wildlife and the continent's biodiversity."[26]

4. Wildlife can be Killed Only for a Legitimate Purpose

This principle legitimizes killing wildlife for "food, fur, self-defense or property protection" and then quickly goes on to distinguish sport hunters—articulation of the attributes of a good sportsman, the ethical code, and the adherence to the rules of fair chase. Pioneering conservationist George Bird Grinnell is credited with the presentation of the "Code of the Sportsman" (see below under "NAM's Ethics"). Accordingly, a sportsman is seen as one who, when hunting, "does so primarily for the pursuit or chase; affords game a sporting chance (fair chase); seeks knowledge of nature and the habits of animals; derives no financial profit from game killed; will inflict no unnecessary pain or suffering on game; and will not waste any game that is killed." In addition, an "aesthetic appreciation of the whole context of sport that includes a commitment to its perpetuation" completes the image of a 'true' sportsman. Perhaps in an attempt to make the NAM relevant to today's society in light of changing public values and increased enthusiasm for plant-based diets, along with declining hunters, promoters praise the NAM for providing "access to massive wild meat harvests" and they applaud the NAM's safeguarding "the right of citizens to access wild protein."[27]

5. Wildlife is Considered an International Resource

Celebrated as another milestone in the history of wildlife conservation, this tenet recognizes that many wildlife species are of international importance, and that management of wildlife is an issue of international concern. An example to note is the Migratory Bird Act, established in 1916.

The need to protect migratory songbirds that protect agricultural crops against insect pests was first recognized by Canadian etymologist, C. Gordon Hewitt, who is honored as the mastermind of the treaty. As The Wildlife Society Technical Review of the North American Model of Wildlife Conservation explains, the historical development of this principle includes the recognition that conserving waterfowl populations requires coordinated and centralized regulations as well as the understanding that international commerce can affect a species' status.[28] These treaties represent agreements between states or nations about the recreational shooting of migratory birds—for example the agreement between the USA and Mexico sanctioning more than 7 million doves being shot out of the sky annually when they attempt to migrate to Mexico.[29]

6. Science is the Proper Tool to Discharge Wildlife Policy

The title of this principle reflects Aldo Leopold's *Game Management*, adopted from Theodore Roosevelt, whose ideas in 1910 came to be known as the Roosevelt Conservation Doctrine. This element mandates that science should inform the decision-making about wildlife management.[30] Mahoney and Jackson note that science was identified early on in the evolution of the NAM as crucial to safeguard wildlife and its sustainable use, even though it was Leopold's writing that set in motion the profession of wildlife scientist and the discipline of wildlife management. Mahoney and Jackson view science as a "safeguard against the politicization of wildlife policy."[31] However, the political take-over of wildlife management in western US states in recent years, shows that their confidence is unfounded (these developments will be discussed in Chap. 6).

7. Democracy of Hunting is Standard

Aldo Leopold called this idea the "democracy of sport"[32] reflecting the NAM's inherent focus on hunting as a democratic process. Often termed "opportunity for all,"[33] this tenet specifically echoes the opinions of Roosevelt, the foremost spokesman for egalitarian allocations, who was prolific in his writings about the societal benefits to be gained by keeping land available for hunting, and by the participation of the 'common man' in hunting. Mahoney and Jackson (2013) expound that the NAM's focus on hunting is a consequence of the historic leadership of hunters and anglers. They further explain that the concepts of fair chase and sport, both borrowed from the otherwise despised[34] European Model, and incorporated in the North American Model, turned hunting into a democratic process, where "everyone has a right and a responsibility for wildlife."[35] It may have become apparent to the reader that NAM promoters seemingly address sportsmen as "the public" as they equate hunting with a democratic process and not the democratic deliberation about hunting that would be done by the entire American public. This is a particularly worrisome concept of democracy given that, as Chap. 2 has shown, the majority of Americans do not hunt or trap, neither want to nor necessarily approve of the killing of wild animals for fun and trophies.

The NAM's Ethics

Both the advancement of the NAM's ethics and its science of wildlife management, with the associated rise of sport hunting, are situated within the larger conservation movement of the late 19th century, which is generally understood to have two streams of thought: the exclusive human-centric philosophy of conservation with its focus on the *utilization* of natural resources, including wildlife, and the aesthetic orientation of the "preservationists" who favored *protection* of nature, and of whom naturalist John Muir was a fiercely committed representative. Prior to Muir, earlier American writers such as Ralph Waldo Emerson in *Nature* (1836) and Henry David Thoreau (1854) in *Walden* had refined the Arcadian Ideal into the more formal philosophy of *Romantic Transcendentalism.* According to this view, the best use of nature was not resource extraction as commodities for human use and material goals, but the appreciation of nature's "intrinsic values and aesthetic qualities through which the human spirit was transformed, a place where one could draw near to and commune with God."[36]

However, Roosevelt's appointment of fellow BCC member Gifford Pinchot (1865–1946), as the first head of the newly established national Forest Service in the United States put a decisive end to *protection* as an official governmental ethos and management goal. Where preservationists such as John Muir had hoped to establish governmental lands as "locked reserves" where forests and wildlife would not be exploited for personal gain, Pinchot championed the Resource Conservation Ethic. This defined scientific conservation through "sustained yield," driven by an extreme anthropocentric utilitarian purpose. Adopted by federal agencies that manage public lands and resources, the essence of Pinchot's philosophy included optimizing the development of natural resources such as timber, "The object of our forest policy is not to preserve the forests because they are beautiful… or because they are refuges for the wild creatures of the wilderness. The forests are to be used by man. Every other consideration comes secondary."[37]

The Resource Conservation Ethic (RCE) rests on two intellectual pillars (equity and efficiency), which stipulated resources to be justly and fairly (democratically) distributed among present and future human generations, and that resources should be used sustainably, and not wastefully. The latter requirement is a direct result from the German influence in American conservation, specifically forestry that favored science, techniques, and efficient production (the forte of Germans). According to Pinchot, a forest had no intrinsic value but rather, its primary purpose was the production of timber. He also promulgated the conviction that conservation had nothing to do with preservation or protection, but everything to do with "wise and efficient use of natural resources, informed by scientific study and practice, with the goal of controlling nature to meet human needs over the long term."[38] Pinchot's ideology was strongly supported by Roosevelt and resolutely shaped wildlife policy and the worldview of the NAM's tenets. Overall, it was the conviction shared by Grinnell and Roosevelt (continuing the white colonialists' beliefs) that "America and Canada were strong because their people had carved the country from a wilderness frontier with self-reliance and pioneer skills. With the demise of the frontier and a growing urban populace, however, they feared that America would lose this edge. They believed that citizens could cultivate traditional outdoor skills and a sense of fair play through sport hunting, thereby maintaining the character of the nation."[39] This male, hunter-centric ethical framework quickly rose to form the dominant philosophical context for the scientific approach, driving conservation for government agencies and the development of principles that eventually led to the formulation of the North American Model for Wildlife Conservation.

The History of Ethics of the 'Sportsman' (Part of Tenet #4)

American historian John Reiger (2001) noted that before the Civil War (1861–1865), the prevalent view of many Americans was that hunting and fishing were activities conducted out of dependence, either to procure food or money. At that time, "if hunting and fishing were spoken of as 'sports,' they were lumped together with diversions typified by chance and a purse, like horse racing, boxing, and cock fighting" (p. 45). This negative image did not sit well with recreational hunters, who thus felt the need to improve the image of hunting by distinguishing themselves from commercial hunters and "pot hunters," a struggle accompanied by the occasional fatality among the two groups. Reiger (1975) writes: "… sportsmen continually complain of arriving at their favorite grounds only to find that all game had been killed or driven out by commercial hunters and fishermen. Until Sportsmen finally defeated the market men in the first quarter of this century and closed off the sale of game for all except for some food fishes, this controversy between the two groups frequently approached a state of war."[40]

Emulating the British characteristics of proper hunting and fishing and the style important to nobility, American sportsmen attempted to not only distinguish, but elevate themselves from their lower counterparts, the market or "pot" hunters. Indeed, sport hunting in the 19th century had assumed the role of "ritualized display of justifying male dominance over subordinates, such as women, boys, immigrants, the lower classes and (former) slaves. Thanks to the efforts of many influential, mostly male, British aristocrats who had immigrated to America, sport hunting had become a worldview, even a religion among hunters.[41] Crafting a dashing, adventurous and honorable image for the wealthy, recreational sportsmen to set them apart from the common market hunter, national periodicals like the *American Sportsmen, Forest and Stream,* and *Field and Stream* defined the ethics and responsibilities of more noble sportsmen. This elevated followers to the status of gentlemen, members of an exclusive club. A class system developed, despite the NAM's principle of democracy regarding hunting standards. The upper echelon of hunters, including trophy hunters, were distinct from the lower order, those who hunted primarily for markets and food. Reiger (2001) notes a typical example of this approach in quoting an editorial by Frank Forster, in which he explains the concept of a 'genuine sportsman:'

> It is not the mere killing of numbers, much less in the mere killing at all; it is not in the value of the things killed, though it is not sportsmanship, but butchery and wanton cruelty to kill animals, which are valueless [as food] and out of season; it is not in the inevitable certainty of success—for certainty destroys the excitement, which is the soul of sport—but it is in the vigor, science [correct technique], and manhood displayed—in the difficulties to be overcome, in the pleasurable anxiety for success, and the uncertainty of it, and lastly in the true spirit, the style, the dash, the handsome way of doing what is to be done, and above all, in the unalterable *love of fair play*, that first thought of the genuine sportsman, that true sportsmanship consists.[42]

Self-restraint is often emphasized in the context of the sportsmen's code. According to a historic analysis by the late ecofeminist and animal rights advocate Marti Kheel, the thinking here was that "by imposing limits on oneself, one achieves the greatest spiritual value, known only to the hunter and God … "The good sportsman moves beyond reliance upon external forms of authority to engage in a personal communion with God, manifested as his conscience."[43] With no one present to monitor sportsmen and their behavior in the woods, this means for wild animals that how they die at the hands of sportsmen depends on the hunter's or trapper's inner moral compass.

The NAM's Science

In his 1933 landmark publication *Game Management*, Aldo Leopold described game management as "the art of making land produce sustained annual crops of wild game for recreational use." A few paragraphs later, Leopold explains that in order to maintain or enhance the yield, management produces a crop which requires the restriction of hunting, predator control, reservation of game lands in the forms of parks, forests and refuges, artificial replenishment (such as restocking and game farming), and the control of environmental factors, including control of food, cover, and disease.[44] Leopold's agricultural view paved the way for a theoretical grounding of the NAM's science, namely the harvest theory, with its objective to achieve a "maximum sustainable yield" the mathematical idea of how much surplus production can be removed while protecting some of the natural capital."[45]

Leopold asserted the importance of including three aspects of the Roosevelt Doctrine, which determined the subsequent course of game management by recognizing that a) all outdoor resources are one integrated whole, b) the conservation of these resources through wise use is a public responsibility and a public trust, and c) science is a tool for discharging that responsibility. Leopold's opus was a response to the need by governmental conservation agencies for scientific information on which to base hunting policies and to improve habitat. An early advocate for the agricultural paradigm, and deeply influenced by Pinchot, Leopold was convinced that game could be managed on a sustained yield basis just like the United States Forest Service manages timber as a natural resource, and deer, just like crops, could be sustainably harvested in a similar way. Having reduced sentient wild animals to nothing more than 'crops' to be harvested like a field of wheat, Leopold's agricultural views paved the way for a science that only considers the species and not individuals even though the violence and death are experienced by the individual animal and only subsequently affects their populations or species.

The Conservation-Industrial Complex

The term "Conservation-Industrial Complex" echoes the "Military-Industrial Complex," a term coined by former US President Eisenhower in his farewell speech in 1961. Eisenhower warned the nation about the power of this emerging complex, referring to the close relations between the government's armed forces and the corporations that supply them.

According to Eisenhower, the complex was not only leading to industry control of the government and economy, but presented a serious threat to democratic governments.[46]

Nearly three decades later, Dutch anthropologist and philosopher Barbara Noske[47] introduced the concept of the "Animal-Industrial Complex," a collective term to describe the multitude of traditions, institutions, and corporate industries which are forcibly turning nonhuman animals into products and services for human consumption. Sociologist Richard Twine has offered the following description of a complex as a "…partially opaque network of relations between governments, public and private science, and the corporate (agricultural) conservation sector. Within the three nodes of the complex are multiple intersecting levels and it is sustained by an ideology that naturalizes the human as a consumer of other animals. It encompasses an extraordinary wide range of practices, technologies, identities and markets."[48]

The fact that these complexes are embedded in the capitalistic fabric of our society explains to a large degree the enormous scope and scale of the exploitation of nonhuman animals. Here, the massive killing of animals is part of an inadequate ethical framework, whereby only human interests in a political economy and capitalism shape human-animal relations and normalize the brutal treatment of animals. Corporations and their financial interests play a central driving force in expanding and intensifying exploitative industries. Twine points to corporate capital accumulation as a significant factor in the emergence of globalized industrial animal productions. Using technology and marketing strategies, corporations help create a socially constructed human identity such as "I am a meat-eater," or "I am a 'sportsman."

NAM promoter Shane Mahoney boasts about the crucial importance of hunting, proclaiming, "In the end, there remains the unalterable truth that life continues as a death-dependent process." Continuing his mantra that implies humans' very survival depends on hunting, Mahoney reveals the staggering economic benefit of hunting in the U.S., listing $25 billion in retail sales, $6.4 billion on hunting equipment, $17 billion in wages, $8.4 billion on "big-ticket" items, and more. He expands the number with "economic multipliers that may be used" to add up to a whopping $86.9 billion that hunting gives the U.S. economy a year.[49]

Whether Mahoney is embellishing these numbers or not, the dark side of it is, what Fish and Wildlife Service agent Dave Hall said in 1991, "We have a war going on, … And as long as enormous profits are attached to wildlife, wildlife loses" (Poten 1991, p. 112). We need to recognize that the hunting industry has grown exponentially since then. With constant proselytizing that the industry of hunting and other animal exploitation industries are essential to life itself, The Animal-Industrial Complex, as well as the "Conservation-Industrial Complex" have placed boundaries around them that are difficult to penetrate and break. With respect to the latter, Colorado-based wildlife researcher Bruce Gill (2004) provided the conceptual framework of the "iron triangle" to describe the near inaccessibility of non-traditional stakeholders, namely citizens who do not hunt or trap, to decision-making in wildlife conservation policy. A group of conservationists recently expanded Gill's concept and following the inclusion of the hunting/shooting-related industry, they coined the terms "Iron Quadrangle" or "Conservation-Industrial Complex." The authors see the following entities as composing this complex and thereby imposing protective barriers: state wildlife agencies, sportsmen (primarily hunters), manufacturers of hunting/shooting [and trapping] equipment and associated paraphernalia as well as policymakers such as the Congressional Sportsmen's Caucus, with congressmen as members, whose fundraising arm—the Congressional Sportsmen's Foundation—is an organization in Washington, DC and perhaps NAM's most powerful lobbying force. With bipartisan representation in nearly all 50 states, its three caucuses—congressional, gubernatorial and state legislative—serve as a network of pro-sportsmen leaders who protect the interests of sportsmen through policies that advance hunting, recreational angling and trapping. Through their support of the NAM and established funding mechanisms, they pledge to "safeguard abundant opportunities for Americans to hunt and harvest wildlife." Because of this interdependency, state wildlife agencies are not only beholden to the interests of sportsmen, but also to the industries and even to not-for-profit organizations linked to hunting and trapping such as The Wildlife Society (TWS) and those whose mission it is to exclusively promote the killing of wild animals. Overall, it is this network of powerful forces that vociferously defends the Conservation-Industrial Complex every time the public tries to curtail or end certain hunting practices and trapping.

Key Players Engaged in NAM Propaganda, Defending (Trophy) Hunting and Trapping

NAM Propaganda by Trophy Hunting Organizations

Hunters have killed a minimum of 1.7 million 'trophy' animals between 2004 and 2014, including at least 200,000 animals threatened with extinction.[50] The website of Texas-based Dallas Safari Club (DSC) prominently features Facebook CEO Mark Zuckerberg and Texas multimillionaire Corey Knowlton, who paid $350,000 in 2014 for a hunting trip to Namibia to kill a black rhino, a member of a critically endangered species as of May 2020.[51] The brutality of killing wild animals for thrill and trophies done by wealthy sport hunters who pay handsomely for the opportunities, came to the public's attention in 2015 through Cecil, a famous and beloved African lion. Cecil was first wounded with a bow and arrow shot by American dentist Walter Palmer, a trophy hunter and member of the Safari Club International (SCI/First for Hunters). Cecil was still choking on his own blood before he died 11 hours later. Two years later, Cecil's son Xanda was also shot by trophy hunters when he too wandered outside a protected zone in Zimbabwe.

In full support of the NAM, the Safari Club International (SCI) is the world's largest trophy hunting organization, with approximately 50,000 members and 200 chapters representing all 50 states in the U.S. as well as 106 other countries. Since 2000, SCI has spent $140 million to protect the "freedom to hunt" by influencing legislation, through lobbying efforts and with donations to political candidates (primarily Republicans) to help elect candidates who support hunting and other interests to sportsmen. SCI gives out awards to those hunters who kill the 'Big Five' (African elephant, lion, leopard, rhinoceros, and Cape buffalo), and has approximately 80 other awards including one that requires hunters to kill hundreds of animals on each and every continent of the planet.[52]

The lives of hundreds of wild animals are auctioned off at SCI's annual conventions. SCI is also heavily involved in influencing international entities, such as the Convention on Trade of Endangered Species (CITES), which regulates trade in protected species of wildlife—including the shipment of hunting trophies. SCI's in-house group of international attorneys is dedicated to defending hunting as a tool for conservation and the NAM. They have relentlessly and aggressively been undermining any efforts to protect wild animals, for example banning the import of body parts (trophies) from African elephants and lions killed by trophy hunters, and their relentless litigation efforts proved successful when wolves lost their ESA protection in 2011. Since 2016, the Safari Club International Foundation has been a member of the International Union for the Protection of Nature (IUCN), the global authoritative body for species status and biodiversity conservation, which maintains the Red List of Threatened Species that assesses species and other ecosystem status. Not only has SCIF worked to ensure that hunting has become a part of IUCN's priorities in the future, but their efforts have resulted in a briefing that presents a favorable position by the IUCN on trophy hunting and the North American state wildlife management system; highlighting the importance of the sale/auction of trophy hunt permits.

UK-based wildlife campaigner and author, Eduardo Gonçalves states that the global dimension and power of the trophy hunting industry are astonishing. He calls trophy hunting "a violent, senseless slaughter on an industrial scale .... whereby "an animal is killed by a trophy hunter every 3 minutes." Trophy hunting groups exploit legal loopholes, which allow them to shoot even the most endangered animals on the planet, just for fun and entertainment. The financial wealth and lobbying power of the industry along with their ways of infiltrating international conservation entities is also pernicious. Having already pushed many animal species to the brink of extinction, Gonçalves warns that trophy hunting could push many more species off the cliff.[53] This is yet another reason why trophy hunting must be banned.

NAM Propaganda by Wildlife Professionals

Shane Mahoney

As a major international trophy hunting organization and in the business of influencing public opinion as well as supporting litigation to protect the rights of sportsmen and sportswomen, the Dallas Safari Club provides a grant to NAM expert, Shane Mahoney. They also feature his video on their website, in which Mahoney talks about the importance of trophy hunting nationally (USA) and internationally. Their financial support extends to "Conservation Force," the Louisiana-based nonprofit organization that Mahoney and John Jackson III codirect that promotes rights for hunters.

John Jackson III

An ex-president of the Safari Club International and a co-author with Mahoney of *Enshrining Hunting as a Foundation for Conservation—the North American Model*, Jackson III has expressed his passion for trophy hunting by stating, "Nothing else in life is more satisfying than an elephant hunt."[54] Jackson's bio posted on Conservation Force's website lists his 'accomplishments' as an attorney in service to the trophy hunting industry. They include establishing numerous trophy hunting sites in African countries and elsewhere, offering the high-dollar opportunity for trophy hunters to safari the world killing wild animals on every continent, including African elephant, lion, leopard, zebra, giraffe, polar bear, white rhino, black rhino, mountain lion, all subspecies of goat, bear and sheep, wolverine, kangaroo and many more. Throughout his life-long career as a 'wildlife attorney,' Jackson has worked tirelessly to prevent the protection of wild animals—especially those animals that trophy hunters are interested in killing—by either defeating proposals to list species under the ESA, or by downgrading the status of other species, and by securing the continuation of import permits for animal body parts (trophies). Ostensibly, the protection of wild animals is not desired by trophy hunters, as it leads to a restriction of hunting opportunities. Conservation Force's website is rather blunt about this and with respect to efforts to protect lions, proclaims "Kenya's proposal to list the African lion on Appendix I is, of course, the greatest threat to the hunting community... We have expended every possible effort to defeat this proposal ..."[55] NAM supporters and trophy hunters are more interested in preserving sport hunting than the survival of animal species, and their behavior indicates a clear lack of regard for the lives and fate of individual wild animals.

Conservation Force

Not only has Conservation Force been granted observer status in CITES, but Jackson is a member of the lion expert committee of the International Union for the Conservation of Nature (IUCN), the world's largest organization that determines the extinction risks of wildlife species, including African lions. Jackson has no scientific expertise but has killed many African lions for trophies and is quoted in a 2020 book exposing the trophy hunting industry as stating, "I can plainly see the African lion that has leaped into the air the moment its head snaps backwards and explodes with smoke from my bullet."[56]

The International Union for the Conservation of Nature is also engaged in efforts to promote Mahoney's "North American Wild Harvest Initiative" (WHI), which seeks to replace or augment meat consumption from the 'livestock' industry with wild animals killed by recreational hunting, thereby helping to boost hunter numbers and make killing wild animals for recreation relevant to food insecurity for many humans around the globe. IUCN's website notes that through his private company Conservation Visions Inc., Mahoney has already secured the support and funding for the study from state agencies, conservation organizations, wildlife management bodies, and hunting organizations.[57] This comes at a time when 'meat' consumption is reaching astronomical levels, where billions of sentient pigs, chicken, cows, sheep, their children (piglets, chicks, calves and lambs) and other 'livestock' are sent to slaughter every year, in the US alone. The industrialized meat industry, with its large-scale factory operations and slaughterhouses, are identified as the largest contributor to climate change.[58]

NAM and the Exotic Animal Hunting Industry

"Ranching for Restoration," is a program by Conservation Force that helps to establish ranches where "exotic, endangered game species" are bred and shot by trophy hunters. Ten percent of the trophy fees goes to the "enhancement" of the species, a mandatory requirement of the Endangered Species Act. Conservation Force is a recipient of the funds from these canned hunt ranches and lists as one of the restoration benefits of this program that it, "Improves the image of the exotic game hunting industry"[59] in North America and elsewhere. Explaining the link between trophy hunting, game ranches and conservation, its founder, John Jackson III states, "At Conservation Force, we want hunting to pay for the conservation of species.... That's what the hunting of super exotics in Texas is doing. It's ensuring that we still have Eld's deer and other animals, not only on game ranches, but also in their native lands." This scheme is called the U.S. Fish and Wildlife Service (FWS)'s "pay to play." The FWS issues kill permits to canned hunting ranches in the U.S. on the basis that 10% of the profits from each animal's death goes to, for example, Conservation Force, which claims to then donate to projects that allegedly support the species' survival in their native homelands. The Ox Ranch in Texas 'offers' 60+ animal species for the trophy hunter to shoot. It's a lucrative business. For an endangered Markhor goat, the cost is $450 per inch of the horn, which ranges between 40 and 49 inches. The price tag for a Safari Club International trophy red stag, up to $35,000.[60]

NAM Propaganda by Wildlife Associations

The Wildlife Society (TWS), founded in 1937, consisting of more than 11,000 members mostly within the wildlife 'profession' lends scientific cover for the North American Model. The Boone and Crockett Club is a leading sponsor of this organization. Mahoney has been their official international representative since 2008, and they also promote Mahoney's Wild Harvest Initiative. The TWS's firearm position statement expresses concern that without secure gun rights, hunting and the NAM's funding mechanism would be severely compromised.[61] This apparently is unacceptable to TWS, and thus they oppose legal efforts that may impact ownership and use of firearms related to hunting and target shooting.

This position places the NAM and The Wildlife Society into the USA national gun control and gun violence debate. Furthermore, Mahoney, a proclaimed leader in wildlife science, management and conservation, pledges his support for secure gun rights insisting that the use of guns is required for wildlife management, and for public hunts, culling operations, other methods along with shooting wild animals for damage control. The Wildlife Society's outright support for recreationally killing wild animals, along with their expressed leading role in defending and promoting the NAM, certainly challenges their claim of being scientifically grounded. Finally, their support for trapping of wild animals for their fur in the name of sustainable use of natural resources, lifestyle and management, as well as their view of animal rights being incompatible with science-based wildlife conservation and management casts doubt over its alleged 'scientific objectivity,' and makes them look more like an advocacy organization for lethal wildlife management and certainly identifies them as a strong anthropocentric and propagandist organization.

NAM Propaganda Through the National Rifle Association (NRA)

In a 2019 newsletter, the Safari Club International/First for Hunters promoted its president Paul Babaz for a position on the board of directors of the National Rifle Association (NRA), a USA-based gun rights advocacy organization with five million members. Babaz described the alignment of interests between SCI and the NRA against the anti-hunting and anti-gun movements, in order "to protect the freedom to hunt whatever game we choose to hunt as well as to protect the right to own the firearms of our choice."[62]

A strong supporter of the NAM, the NRA was also a sponsor of the Boone and Crockett Club's 30th Big Game Awards Benefit Auction in 2019. At this event, several hunting trips were auctioned off to shoot goats, deer, and predators, including bear and mountain lion; a 2-week-long trophy hunt to kill two wolves in British Columbia was sold for $5750.00.[63] There is a long historic relationship between the NRA and trophy hunting organizations; "Few people know this, but there was a time the B&C records books were published by the NRA," said Boone and Crockett Club President Timothy C. Brady. "What's more important to know is that the North American Model of Wildlife Conservation could not and would not have happened without the right of citizens to bear arms, including the use of firearms in hunting. When we think of standing up for this constitutional right, we think of the NRA."[64]

NAM Propaganda Through the National Trappers Association (NTA)

Represented through chapters across the United States, the National Trappers Association (NTA) is "committed to defending and promoting the safe and ethical harvest of furbearing mammals and to the preservation and enhancement of their habitats."[65] A 2009 article quotes then NTA president Kraig Kaatz, "The National Trappers Association began 50 years ago to combat anti-trapping legislation. Today, that focus continues, only now anti-trappers are also using the court system and ballot initiatives to try and end trapping."[66] Indeed, as their website states, through their attorneys and lobbyists the NTA is heavily engaged in fighting legislative bills they perceive to negatively impact their trapping activities, and aggressively pursuing those bills that would expand their "American heritage" of trapping for recreation. Trappers go to great lengths to protect trapping, and continuously opposed protecting Canada Lynx (now listed as endangered) in 2015[67] and opposed stronger protections for wolverine in 2013–2014, even though there are less than 300 wolverines left in the U.S. The fight by wildlife advocates for ESA protection of wolverines, which began in 2000 continues as of 2021.[68] The North American Model is prominently featured in the 2005 Trapper Education Manual, developed by the International Association of Fish and Wildlife Agencies (IAFWA). Students are encouraged to learn about the NAM to understand the legitimacy of trapping. The Manual even adds trapping to tenet #7, calling it "Democracy of Hunting and Trapping"[69]

NAM Propaganda Through National Trophy Hunting Organizations

The leading role of the Boone and Crockett Club (BCC) in the formulation of the NAM has already been described above. The BCC makes clear its focus to maintain game species for hunters by killing wolves. They boast that one of their studies on predator/prey relationship caused the timber wolf to lose ESA protection, or in the BCC's words, they "helped remove the Eastern timber wolf from endangered status while changing the public's perception of wolves."[70] In 2000, the Boone and Crocket Club organized a summit for 35 organizations representing 4.5 million pro-hunting individuals, including the Safari Club International, the Rocky Mountain Elk Foundation (RMEF) and other influential trophy hunting organizations, along with the National Trappers Association (NTA). Under the tagline *Action: What's Next*, they state, "The time is right for wildlife management advocates to be a larger voice in how government and the Congress deal with wildlife and public access to it in America."[71] The state wildlife agencies seemed to have followed this advice when the Association of Fish and Wildlife Agencies (AFWA) officially endorsed the NAM in 2002.

Using the headline "Hunting is Conservation," the Rocky Mountain Elk Foundation, also based in Missoula, Montana, works to ensure sportsmen's hunting heritage. Their website claims that hunters were the first conservationists and that they have remained the "mightiest force" of the conservation world. The RMEF fights an outright war against wolves, supporting any bill and move that could lead to more wolves being killed. For example, RMEF was in strong support of the delisting of gray wolves from the ESA so that wolves could be 'managed' by the states through hunting and trapping.[72] To prove the length these sportsmen are willing to go in order to kill wolves, their former leader, David Allen, once stated, "To keep wolf populations controlled, states will have to hold hunts, shoot wolves from the air and gas their dens."[73] RMEF has donated $25,000 to the Foundation for Wildlife Management, an Idaho-based nonprofit organization that pays bounties for each wolf killed by a hunter or trapper. Under the tag line, "Tired of wolves interrupting your hunt and displacing your wildlife?" one reads that the Foundation for Wildlife Management (F4WM) has 'removed' over 1100 wolves, actions made possible by $750,000 in private funding from sportsmen, ranchers, grants, and individual citizens. F4WM spent well over $200,000 in bounties, reimbursing the killing of 271 wolves in the 2019/2020 season alone.[74] Now it is legal to kill wolves year-round in Idaho, with the goal of eradicating 90% of the population. As of this writing in February, 2022, 500 wolves have been killed in seven months. All eleven board members of F4WM are wolf trappers and hunters receiving bounty money, which rose to $2500 per wolf in Idaho. Several officers were caught appropriating donations for their own use as well. F4WM has expanded into Montana, holding fundraisers, and offering $500 per dead wolf. Montana's governor, Greg Gianforte, illegally shot a trapped wolf with a Yellowstone collar ten miles north of the park. He paid no fine and received no penalty. The wolf is mounted in his office.

Brands/Corporate Supporters of the NAM Include:

Walmart/Multinational Retailer; Cabela's/Hunting and Outdoor products; Hornaday/rifles, ammunition; Kenetrek Boots/hunting, hiking, working boots; Scheels/outdoor gear; Bass Pro Shops/hunting, outdoor gear; Global + rescue, Ripcord travel insurance; USA/Midway/hunting and outdoor gear; Wildlife Partners/ranching services for exotic shooting industry; Capital Farm Credit/Finances; Richard Childress Racing/Sportscar racing; Shimano/biking, boating, fishing; Vista Outdoor/Shooting sports and Outdoor products; Black Rifle/coffee; Reynolds American Inc./Tobacco products; General Dynamics/Global Aerospace and Defense Company; Shell/Oil company; Anheuser-Busch/Budweiser beer; Exxon Mobil/Oil company; Sempra Energy/utilities and infrastructure; Altria/Tobacco/Alcohol and Cannabis; Charter Communications/Spectrum/Grady-White Boats/Boat Company; Luvata/Copper products' The Sportsman's Warehouse/hunting, trapping and outdoor products; a long list of rifle, ammunition, outfitters, finance and oil companies, trophy hunting and exotic shooting ranches, and other partners, including The Nature Conservancy.

Concluding Comments

The themes that have emerged in this chapter bring into question the relevance of the North American Model for thinking about wildlife ethics in contemporary society, and for wildlife conservation measures in particular. The list of NAM's elements encapsulating the human/sportsmen's relationship with wild animals is characterized by a cold and detached attitude that entirely ignores the wellbeing of individual animals. Here, wildlife conservation and the process of democracy has nothing to do with a public's right to intervene and protect wild animals, or a wild animal's right to life. Rather, conservation under the NAM promotes the belief in a human's inalienable right to hunt, trap, and kill sentient and unsuspecting beings. The NAM and its ethics and science reflect an old and untenable paradigm that has not kept pace with developments in social ethics and public concerns. Sportsmen kill intelligent and sentient wild animals who matter to themselves and to others and are embedded in a tightly knitted network of social and ecological relationships. Sportsmen not only destroy animal bodies but also animal minds.

The tenets of the NAM are preoccupied with a focus on the continuation of recreational hunting of game animals. It suffers from an inherent contradiction of claiming opposition to market hunting, while sanctioning commercial trapping and the fur trade, which commercializes wildlife as being a product for a market. Pertinently, this contradiction comes to light when one scrutinizes trapping through the filter of the sportsmen's ethics code, one of the key concepts of the NAM. Not only does recreational and commercial trapping not meet the "gentleman-like" way of killing animals, but by approving fur markets, the NAM contributes to the overall existence of wildlife markets (legal and illegal), which pose a significant threat to the survival of many species. Finally, the NAM categorically opposes game farms while not showing any concerns about fur farms, another contradiction apparent in the NAM. Overall, the NAM appears to be significantly linked with the identity of the sportsman and aimed at protecting sportsmen's activities. This focus leads to yet another contradiction, where the NAM's survival with its core concept of sport hunting is recognized as being dependent upon the public trust doctrine, which declares wildlife a public trust; however, it is nearly impossible for non-hunting and non-trapping wildlife supporters to restrict or ban hunting or trapping when deemed unacceptable. The NAM's leading role in the disastrous global trophy hunting industry identifies this as an ecologically catastrophic, socially and ethically repugnant approach to conservation that seeks to protect hunting, not wild animals and their homes. To conclude, the NAM is ill-equipped to prepare conservationists and the public to address the most pressing threats to wild animals—human population growth, habitat loss/fragmentation, and global climate change. The future of conservation can and should be guided by a set of principles and tenets that address these contemporary, global, and ethical concerns along with its root cause of a lived human supremacy worldview. With sport and trophy hunting, fishing, and trapping as its focus, the NAM does not provide any inspiration for an ethical relationship between humans and other species.

Notes

1. White, et al. "Trapping."
2. Bull, "Jekyll & Hyde?"
3. Kheel, *Nature Ethics,* 70.
4. Boone and Crockett Club, "Mission."
5. Ibid, "North American Model of Wildlife Conservation."
6. Reiger, "*American Sportsmen.*"
7. Boone and Crockett Club, "North American Model."
8. Organ, "Born."
9. Kheel, *Nature Ethics.*
10. Boone and Crocket Club, "North American Model."
11. Geist, et al. "Hunting Has Defined the North American Model," 175.
12. Mahoney and Geist, "The North American Model," 1.
13. Miller, "Hunting as a Wildlife Management Tool," 8.
14. Conservation Visions, [DVD], "Opportunity."
15. TWS, "Baiting."
16. Sullivan's, "Carcass Disposal."
17. Elliot, "Mahoney Advises Hunters," 10B.
18. Realtree, "Female Shooting Sports."
19. While the numbering and ordering, and even the titles of the NAM's elements often vary, I use those listed in the technical review provided by Organ et al. (2012) since this is an update of the original version by Geist et al. (2001).
20. Horner, "Embryo."
21. Sax, "Effective Judicial Intervention."
22. Serfass et al., "North American Model."
23. Mahoney and Jackson, "Enshrining Hunting," 452.
24. Organ, et al. "Technical Review," viii.

25. Mahoney and Jackson, "Enshrining Hunting."
26. Geist, "Means of Creating Wealth," 289.
27. Mahoney and Geist, *North American Model*, 2–3.
28. Organ et al., "Technical Review."
29. Kulowiec et al., Mourning Dove."
30. Organ et al., "Technical Review."
31. Mahoney and Jackson, "Enshrining Hunting," 456.
32. Meine, *Aldo Leopold*, 169.
33. Conservation Visions, [DVD], "Opportunity."
34. Due to the fact that hunting was historically mostly reserved to royalty and wealthy landowners.
35. Mahoney and Jackson, "Enshrining Hunting," 454.
36. Van Dyke, "History and Distinctions," 12.
37. Graham, *Man's Dominion,*109.
38. Van Dyke, "History and Distinctions," 15.
39. Brands, *Last Romantic*, 24.
40. Reiger, 1975, *American Sportsmen*, 28.
41. Reiger, 2001, *American Sportsmen*, 49.
42. Ibid, 48.
43. Kheel, *Nature Ethics,* 122–123.
44. Leopold, *Game Management*, 9.
45. Lavigne, *Gaining Ground*, 4.
46. NPR, "Eisenhower's Military-Industry Complex Warning."
47. Noske, *Humans and Other Animals.*
48. Twine, "Revealing."
49. Mahoney, "Hunting as Largest Funding Mechanism."
50. IFAW, "Killing for Trophies," 4.
51. IUCN, "African Rhinos."
52. Gonçalves, *Exposed,* 10–11.
53. Gonçalves, "Scale and Power."
54. Gonçalves *Exposed,* 10.
55. Conservation Force, "Will Lion Hunting Survive?"
56. Gonçalves, *Exposed*, 9.
57. IUCN, "Wild Harvest Initiative."
58. Carrington, "Eating Less Meat."
59. Conservation Force, "Ranching for Restoration" (pamphlet), 2.
60. Ox Hunting Ranch, "Red Stag."
61. TWS, "Firearms."

62. Professional Outdoor Media, "Support."
63. Boone and Crockett Club, "30th Awards Auction."
64. Boone and Crockett Club, "BC &C Welcomes NRA.
65. National Trappers Association, "About."
66. Gross, "National Trappers."
67. Associated Press, "Trapping Restrictions."
68. Perkins, "Commissioners."
69. IAFWA, "Trapper Educational Manual," 15.
70. Boone and Crockett Club, "History."
71. Boone and Crockett Club, "Action: What's Next."
72. RMEF, "Strongly Supports."
73. Stalling, "Once-Proud."
74. Foundation for Wildlife Management, "Tired of Wolves," and also add it to the general bibliography as: Foundation for Wildlife Management, "Tired of wolves interrupting your hunt and displacing our wildlife?" Accessed May 2020. https://www.foundationforwildlifemanagement.org/

4

The Existing Critique of the North American Model of Wildlife Conservation

Inadequate History and Ethics

The North American Model of Wildlife Conservation remained largely unchallenged until 2011 when a group of ecologists and ethicists provided the first critical review. Their paper "An Inadequate Construct? North American Model: What's Flawed, What's Missing, What's Needed"[1] set in motion a litany of examinations of the NAM that is ongoing, pointing out the NAM's weaknesses and detrimental impacts, questioning or rejecting the NAM as an acceptable approach to the conservation of wild animals. Several themes have emerged, all centered around the charges against the NAM's hunting-centric approach: its focus on 'sportsmen' while excluding the non-hunting and non-trapping public, its aggressive persecution of wolves and other predators, certain use of language, marketing strategies, and ethics surrounding funding mechanisms. A main criticism of the NAM is that it has been found to empower the old paradigm while hindering new, inclusive, and more effective approaches that respond to the public's broad spectrum of values.

In their initial 2011 critical review, Michael Nelson and colleagues identified several flaws, including the NAM committing a logical fallacy

A. Heister, *Beyond the North American Model of Wildlife Conservation*, The Palgrave Macmillan Animal Ethics Series, https://doi.org/10.1007/978-3-031-14149-2_4

by relying on past behavior to justify future behavior. The authors explained that society would not argue for a future of child slave labor or gender discrimination simply because these practices are part of our history. Related is the concern that NAM proponents may obfuscate their motivations of defending hunting rather than conservation, which is a hallmark of inadequate reasoning. This preference comes to light especially in cases that ignore issues of ecosystem impacts when those issues conflict with the interests of hunters—such as an overabundance of white-tail deer and a hostile attitude toward predators. Furthermore, the authors object to the claim that the NAM's narrative encompasses the entire history of conservation by pointing to other forces from non-consumptive users of nature as important in wildlife conservation practices in North America, concluding that "hunting is not *necessary* for conservation."[2]

Selective Focus on 'Sustainable Use': Protects Sports/Trophy Hunting

According to Mahoney and Jackson (2013), the North American approach of the 'wise use' of wildlife is of incredible importance for the *future of hunting* [emphasis added] because it reflects the vision of hunters like Aldo Leopold and Theodore Roosevelt, who have always maintained the conviction that lawful hunters and their personal engagements with wildlife could represent a powerful force in conservation.[3] This thinking also indicates sportsmen's general affirmation that sportsmen control wildlife and land, a powerful role that they are not willing to retreat from, compromise or give up. Sportsmen thus need the public to accept the concepts of 'sustainable use' as well as 'hunting and trapping are conservation' as truths in order to maintain this type of control. To this end, and to guarantee as best as possible a future for hunting, sportsmen need the public to accept the idea of wild animals as resources for humans to use. Here, the NAM serves as a vehicle to get the public to agree with these ideas and accept that 'there can be no conservation without hunting.'

NAM's Bias on (Huntable) Animal Species, Ignores Others

Some of the NAM's key supporters celebrate its grand achievement as restoring the "continental pyramid of hunted wildlife."[4] The bias towards huntable animal species is one of the major weaknesses of state agencies' policies, in that it comes at the expense of other species that sportsmen are not interested in. Conservationist and author, George Wuerthner explains that agencies often sustain (so-called) invasive species such as wild boars despite their detrimental impact on native animal species (e.g., when these feral pigs prey on native salamanders). In addition, wild animals from other countries (so-called 'exotics) have been taken from other countries and introduced in North America for the sole purpose of adding hunting opportunities. The bias of the NAM and thus governmental agencies towards huntable, terrestrial animal species then comes at the expense of other species and ecosystem and conservation. While there is an ongoing heated debate about nonnative versus native species calling for the eradication of the former in order to protect the latter (but see the discussion of Compassionate Conservation in Chap. 8), under the NAM, many states in North America continue to maintain huntable populations of introduced ungulates (e.g., oryx, barbary sheep in New Mexico). Furthermore, when it comes to fishing opportunities, state wildlife agencies "operate as monopoly industries," notes wildlife protector and New Mexico resident Kevin Bixby. He explains that these agencies have co-opted a public resource—native aquatic ecosystems—in order to produce a consumer product—fishing opportunities for non-native fish—which they then sell to generate revenues for themselves. Bixby uses the state wildlife agency of New Mexico as an example, which "dumps more than 15 million non-native fish into the state's waterways annually, all of them predatory species like rainbow trout and walleye. Some of these naïve captive-raised fish, which frequently don't survive more than a few weeks in the wild because they fall easy prey to human anglers or other predators, have to be obtained from other states to meet perceived demand."[5] Nelson and colleagues have added additional concerns by linking NAM's primary interest of protecting hunting of game species with the neglect of attention paid to threats imposed by commercial fisheries onto fish and other marine lives, which also led to their conclusion that the NAM does not offer any useful insights into wildlife conservation.[6] In the same vein, the late predator ecologist and author, John Laundré, stated that the continued "support by hunters for introducing 'exotic' animals into ecosystems disqualifies hunters from calling themselves, conservationists."[7]

Members of species such as prairie dogs, herbivorous 'rodents' with a complex social structure and sophisticated language, are considered shooting targets and "are often blown away by some to see the red mist of their blood hanging in the sky."[8] This killing comes despite the fact that prairie dogs are considered a "keystone species" because they support 160–200 other animal species, and their burrows provide habitat for many other species, including burrowing owls (*Athene cunicularia*) and black-footed ferrets (*Mustela nigripes*), an extremely endangered species. Labeled a "vermin" by ranchers, these wild beings, who "love to kiss" as the famous American actress and political activist, Jane Fonda learned[9] are not protected and can be killed by anyone at any time in most states.[10] For example, the Montana Department of Fish Wildlife and Parks (FWP) classifies prairie dogs as nongame wildlife, while the Montana Department of Agriculture classifies them as an 'agricultural pest' due to their perceived interference with human enterprises of agriculture and farming (thus, neither classification provide any protection for prairie dogs). Consequently, most public lands, with the exception of a few wildlife refuges in some regions of the United States, are open for anyone to shoot prairie dogs, no license required, no questions asked.

State wildlife agencies go to extreme lengths in order to please the interests of sportsmen and recruit new hunters. For example, from 2021 on, the Montana Department of Fish, Wildlife and Parks plans to spend $1 million to release around 50,000 pen-raised ring-necked pheasants for hunters to shoot. It does not matter that these birds also are not native to Montana, that they have a low survival rate once released, what matters is that their release occurs shortly before the youth hunting season starts, which is part of the R3 program (recruit, retain and reactivate) to entice new (young) hunters into the animal-shooting business.

It is also worth noting that in this specific example, different versions of the previously discussed 'Complex' intersect. For example, as the

Montana example shows, human prisoners had previously been used to raise cows and their calves for milk before the animals were sent to slaughter, and now the prisoners raise ring-necked pheasants, who will get released to be shot by hunters.[11] This case exemplifies an intersection of the Prison/Conservation and the Animal/Industrial complexes.

Contrary to claims by some NAM advocates that non-game species have benefited from the halo effect shone on them simply by sportsmen's focus on game species, many wild animals struggle with decreased levels of security, safety, and survival. The list of threats imposed on them related to an ever-expanding human population and increased economic development is long and includes environmental changes (habitat degradation and loss, pollution, climate change), and broad social changes (economic upheaval, industrialization, energy development, expansion of and persecution by agricultural industry, urbanization) that drive extinction rates. Critics have emphasized that the NAM's "actual conservation record over the past is not as impressive as is sometimes suggested;[12] in September 2021, the US Fish and Wildlife Service (USFWS) took 23 species off the Endangered Species List because they are extinct.[13] Human activities, including fishing, hunting [and trapping] are weakening populations and are in part responsible for ongoing, massive, and forced extinctions of nonhuman animal species. To date, 2244 species are listed as threatened or endangered under the Endangered Species Act, with 1618 of those in the United States alone.[14] An ever-increasing number of reports issue stark warnings of a growing risk of possible ecosystem collapses worldwide.[15] International entities such as the United Nations, are ringing the alarm of an imminent extinction of one million plant and animal species in the next few decades if nothing changes (see Chap. 8).

The fact that neither the NAM's narrative nor its elements address the alarming, ongoing 6th species extinction, the most profound and grotesque, human-caused phenomenon of all times, is conceptually a failure at a minimum. After all, as ecology scholars M. Nils Peterson and Michael Nelson (2016) observe, the sixth great extinction event occurred at the same time and under the same management approach that miraculously saved a few select wildlife species.[16] The authors make clear that they do not see the NAM as a driver associated with the ongoing extinction crisis, but they point toward North America as a place that is experiencing extinctions and threats similar to those seen globally.

Some authors have remarked that the NAM even hinders the development of holistic approaches to conservation. Michael Soulê, the "grandfather of conservation," clearly perceived conservation biology and wildlife management as differing in goals when he founded the Society for Conservation Biology in 1985, in part due to the need for a more holistic approach that would address the entire spectrum of conservation challenges threatening the conservation of biodiversity.[17]

Valerius Geist (2006), who originally conceived of the idea of the NAM has claimed that predators are "better off under hunting regulations, because the kill is closely controlled, is under constant public scrutiny and persons are held accountable for each kill."[18] This thinking conveys a lack of empathy and compassion for the individual animal, let alone consideration of the animals' standpoint. Geist's statement certainly does not reflect reality where wolves, mountain lions and other predators are mercilessly executed by sportsmen and agency personnel alike. Hunters often view predators such as bears, wolves, coyotes and mountain lions as competitors and want them killed; ranchers perceive carnivores as a risk for their financial profiting from cattle, and frequently demand their killing. The United States Department of Agriculture's Wildlife Services usually executes the killings of the (native) carnivores on behalf of the corporate ranching and livestock industry.

As previously noted, Nelson et al. (2011) have noted that when conservation conflicts with hunters' interests, the detrimental impacts of an overabundance of game species such as whitetail deer (intentionally and artificially created through wildlife management) are overlooked while predators are killed and their 'ecosystem services' are ignored.[19] This goes against the best available science that shows that predators such as wolves play vital roles in the functioning of ecosystems. As top carnivores, wolves, mountain lions, bears and coyotes, have found to regulate their own numbers for example "by limiting reproduction to alpha individuals within a group, investing in lengthy parental care, and infanticide." Given their own management ability, hunting is not only not needed to keep populations of top predators in check, but rather hunting has the opposite effect because it disrupts the [animals'] social interactions through which self-regulation is achieved.[20] As Vucetich and Peterson have stated, the claim that hunting is supported by the best available science is a misrepresentation as there is no science that concludes it is *necessary* to hunt wolves. They explain, "As science cannot tell us what is good or bad for society," in the end, "good wildlife management is the judicious balance between science *and* democracy," and they concluded that Michigan's wolf hunting presents a violation of four of the NAM's own tenets: #1. Wildlife is Held in the Public Trust, Principles of Democracy -- Non-Frivolous Use-#5, and Best-Available science.[21]

Rejection of current wolf management practices also came from within ranks of those listed among pioneers of conservation by NAM supporters. In 2012, the family of Olaus J. Murie, who is sometimes called the "Father of Elk Management"[22] and who, along with Aldo Leopold and a handful of other pioneers, was among the first champions of biodiversity and preservation, revoked the use of the Murie name for an annual award given by the Rocky Mountain Elk Foundation in response to their "all-out war against wolves."[23] Indeed, predators in the United States are being persecuted by governmental agencies, and contrary to NAM promoters' averments, they are not better off under the NAM at all. Instead, governmental agencies fight an outright war against these carnivorous species, largely because of the anthropocentric mindset of agency personnel. Their agencies' focus on serving 'sportsmen' rather than wild animals, has not changed since the horrific predator control programs of the past that sought to eradicate these animals on behalf of ranchers and sportsmen. Also, when wildlife-killing programs are implemented, promoted, executed, protected and fiercely defended by the "Conservation Industrial Complex," this creates further conflict between the government and public, and in the process, a worsening of their perception of 'sportsmen' as well as ranchers, on whose behalf wolves and other predators are often killed.

Not a Product of Consensus and Dividing Wildlife Conservation and the Public

In their discussion of the NAM's shortcomings, New York-based environmental and science communication scholar Andrea Feldpausch-Parker et al. (2017) highlighted the fact that the NAM did not emerge out of a consensus achieved within the entire wildlife conservation community; it has not even been sufficiently discussed, let alone scrutinized from within the wildlife conservation community.[24] Instead, the NAM has been, and continues to be, endorsed, promoted and aggressively disseminated by a subgroup of wildlife management professionals along with influential sportsmen's groups and trophy hunting organizations, governmental agencies and associations representing them. Despite the lack of consensus, the Association of Fish and Wildlife Agencies (AFWA) representing state wildlife agencies on all levels in North America and Canada, has endorsed and promotes the NAM. Interestingly but not surprisingly, sponsors of their annual meeting include governmental agencies such as the US Fish and Wildlife Service (USFWS) and the US Department of Agriculture (USDA) as well as The Wildlife Society (TWS) and several sportsmen organizations such as the Boone and Crockett Club, Rocky Mountain Elk Foundation and the Wildlife Management Institute. The backing through the AWFA gives the NAM agenda significant influence in directing sportsmen-driven wildlife management and policies that sanction injuries and premature death of hundreds of millions of defenseless and unsuspecting wild animals every year in North America.

As previously discussed, a historic fracturing of the wildlife conservation community happened when Roosevelt and the Boone and Crockett Club succeeded in overpowering those conservationists as John Muir, who advocated for hunting to play a lesser role in conservation efforts. Canada-based author and Professor Emeritus of Communication, Media and Film, Max Foran (2018) writes, "[T]hrough the first half of the twentieth century, the hunting lobby integrated itself into the conservation movement by influencing game laws, hunting licenses, and the establishment of sanctuaries and wetlands where game could breed and thrive."[25]

The historic and ongoing insertion of the NAM into the already fractured conservation community continues to cause fraction and animosity between groups with opposing perspectives of the role of hunting in conservation. Essentially, by championing a wildlife-conservation system that values the interests of hunters over those of non-hunting wildlife enthusiasts the NAM serves as a polarizing force within wildlife conservation. For example, upon noticing that new recruits into the wildlife profession are increasingly those with a so-called 'non-traditional' wildlife management background (e.g., women, ethnic minorities, urban residents and non-sportsmen) and without a commitment to consumptive use of wildlife, the pro-hunting community mobilized a new program called "Conservation Leaders for Tomorrow."[26] The intent of this program seemingly serves to instruct the non-hunting and non-trapping university students of wildlife-related degrees and natural resource professionals about the importance of hunting. In absence of any ethical discussions and education on other approaches to conservation that do not include hunting and trapping, this type of promotion of the NAM comes close to indoctrination, along with peer pressure onto those who do not want to kill animals. This splitting of human supporters of wild animals into at least two camps in stark opposition to one another has occurred within the wildlife conservation community and among the public. In the process, the NAM continues to exclude and marginalize the interests of the non-hunting and non-trapping wildlife supporters and by doing so, prevents an otherwise potentially broad supporter base wild animals so desperately need.

As a final observation, one cannot help but notice that NAM promoters, ostensibly so concerned about the European model where hunting is the domain of the wealthy and powerful, have turned into just that, an elite—a dominant minority imposing their questionable pastime activity onto others. In the process, the NAM has created a 'class system' among sportsmen whereby following big money leads to the trophy hunting of 'exotic big game' and associated awards and recognition, all reserved for the wealthy and powerful. Down the hierarchy, the killings of the more common animals, including deer and antelope, falls to the 'common' man or woman. In sum, the NAM continues to not only cause a division within wildlife conservation but also among the general public and thus also reduces resources for the protection of nonhuman nature.

NAM Has Established Boundaries: Excludes the 'Non-consumptive' Public

The long involvement of sportsmen in the establishment of creating governmental agencies, laws, regulations and policy related to wild animals has also led to the establishment of the wildlife profession, including the university-trained wildlife biologist or manager. This way, there is an invisible yet clearly palpable boundary around wild animals and their management through the governmental wildlife manager—the self-declared authority.

From a structural perspective, there is a clear boundary where hunters and trappers have political and policy influence and a loud voice to influence wildlife management, while the general public is being kept outside. The state's governor generally appoints the director of the state wildlife agency. The director or head of a state wildlife agency is usually responsible to some type of politically appointed commission, board or council, whose powers typically range from setting hunting, fishing and trapping seasons and kill quota per individual (bag limit) to describing broader management goals and objectives for the state.[27] Commission members are often required to have a hunting/trapping license among other requirements. Nie et al. (2017) explain that "commissions were created so that sport hunters had a voice in preventing a return of widespread market hunting."[28] In Montana, the Governor not only appoints the director of the state wildlife agency but also the individuals to serve on the agency's Commission, which is comprised of seven members. Strikingly, in the 21st century, the Montana FWP Commission overwhelmingly represents commercial and recreational interests and lacks representation of Native American tribes and other members of the public that are not sportsmen. This setup gives sportsmen enormous power in influencing and directing wildlife management (and thus the treatment of wild animals) through hunting and trapping, while the public has no such power. Resources and powers of governmental agencies are being used to promoting hunting and trapping as *the* 'priority use' of wildlife, and the sportsmen-centric commissions, provide access to policymaking to 'sportsmen,' and not for the general public.

Along the same line, the frequent attitude among sportsmen that wildlife management is a 'sportsmen domain,' along with generally dismissive attitude among sportsmen when the non-hunting public requests more inclusivity and involvement in policymaking affecting wild animals has a long history. For example, Valerius Geist, acknowledges that naturalists helped hunters in their efforts to promote conservation, and then emphasizes that "it is hunters however, or, more accurately, *hunting*," that led to the foundation for North American wildlife conservation.[29] Finally, another of Geist's publications further asserts that specific legislation to save endangered species could not have been achieved without an existing hunting culture,[30] a claim that has been refuted by Michael Nelson and colleagues writing that in the 1960s, conservation was led by non-hunters whose efforts culminated in such significant achievements as the U.S. Wilderness Act (1964), Endangered Species Act (1973), Clean Air and Water Acts (1963 and 1972 respectively), and the National Environmental Policy Act (NEPA).[31]

NAM's Narrative Ignores Leading Female Conservationists

While the NAM is preoccupied with a portrayal of historic activism by a small group of powerful, primarily white men to save hunting, there was at the same time grassroots activism par excellence for wildlife protection. One of those leading figures was Rosalie Barrow Edge (1877–1962), a champion of suffrage from New York, animal advocate, and one of the most influential and determined female conservationists. Motivated by disgust about leading conservationists' corruption, and cruelty, Edge established the Emergency Conservation Committee (ECC) in 1929 to expose the conservation establishment's ineffectiveness, and strongly advocated for species preservation and the protection of entire ecosystems instead of the conservation of certain species of interest to hunters, who were at that time predominantly men.[32] To give an example of her activism, when Edge found out that "gunners" had shot hundreds of passing hawks at a ridge in Pennsylvania, she bought the 1400-acre property, banned the hunters, and thus established the first reserve in the world for raptors called 'Hawk Mountain Sanctuary.'[33] This is the same sanctuary that helped Rachel Carson in her discovery of the decline in juvenile raptors was linked with the use of DDT (as demonstrated in her book, *Silent Spring*). Edge's other conservation accomplishments include leading the national grassroots campaigns to create Olympic National Park (Washington, 1938) and Kings Canyon National Park (California,

1940). Even in *The North American Model of Wildlife Conservation*, the most recent 2019 book promoting the NAM, the authors found no place to mention or quote female conservation activists other than Rachel Carson. While the reader of this publication learns about a trapper named Grey Owl, the existence and life of his wife is entirely omitted. Notably, his wife Anahareo (Gertrude Bernard), a free-spirited Canadian (1906–1986)[34] was already a naturalist and animal rights advocate who turned her husband from a trapper into an animal-respecting conservationist. Anahareo campaigned for the banning of leghold traps and poison, and in her later years, was inducted into the Hall of Fame of Animal Rights activists. Certainly, an interesting constellation and occurrence in conservation history, however, this seems of no interest to the NAM authors. The NAM's makes no mention of Rosalie Edge or any other female conservationists of the time, a fact also critically remarked upon by M. Nils Peterson and Michael Nelson (2016), who commented that not even the highly influential Audubon Society, finds mentioning.[35] Indeed, the NAM's narrative features male-run hunting clubs and credits them with "some of North America's most important and enduring conservation legacies,"[36] while ignoring the fact that women such as Harriett Lawrence Hemenway and Minna B. Hall founded the Massachusetts Audubon Society in 1896, whose efforts were indispensable in the passage of the Lacey Act of 1900, which spurred several significant conservation efforts, including the Migratory Bird Treaty of 1916 and the Endangered Species Act. As Peterson and Nelson remark, "[O]ne cannot imagine a framing more centered on white male hunters than that contained in this historical overview of the origins of wildlife conservation in the United States."[37]

A Minority That Uses Power and Ideology to Dominate

The promotion of the NAM relies predominantly on wildlife professionals who are part of the wildlife management establishment, in other words, they are powerful individuals leading the conversations about the

NAM and its ideology of consumptive use. As discourse analysis shows, the mutual relationship between discourse and power lies in the ability of language to control or affect the behaviors and thoughts of others. These strategies are applied for the promotion of the NAM and accomplished through the leadership by wildlife professionals such as Shane Mahoney, who is described as a 'leading NAM expert' and a frequent presenter in videos and at sportsmen's gatherings and conferences, including at the highest international levels. Other NAM promoters include other wildlife professionals and governmental officials, who have the ability to use their authority and influence to guide, lead or control others through speech and conversation. Feldpausch-Parker et al. (2017) have pointed out that NAM promoters are not the dominant group within the wildlife conservation community in North America, yet they are certainly the dominant group "among wildlife *management* professionals that are responsible for managing wildlife for the public trust."[38] These important power imbalances are being used in interactions of governmental agency personnel and the general public, where the former assume the acceptance of hunting and trapping by the general public. "Discourse is power. Power and ideology work in tandem through discourse to legitimate the power of a dominant group in particular circumstances."[39] Having defined conservation through their own, consumptive ideals, despite being a minority within the conservation community, NAM promoters nevertheless exercise dominance through a discourse that maintains the privileging of consumptive use ideology over others. To this end, they promote the NAM's tenets as truths to be adhered to by wildlife professionals and to be disseminated to the public through formal education.[40] Feldpausch-Parker et al. further call attention to the NAM functioning as "a symbol of the underlying connection between conservation and consumptive use that is so important for the persistence of hunting and angling [and trapping] within wildlife management."[41] Another characteristic of reinforcing their ideology is the relentless warnings about sportsmen being indispensable to conservation, which also serves to elevate hunters and trappers over non-consumptive conservation practitioners.[42] Finally, through institutions of authority such as government agencies, The Wildlife Society (TWS), and the use of charismatic and influential individuals such as Shane Mahoney, the consumptive use messages appear to be accepted, natural and grounded in consensus, and they are relentlessly communicated to the public.[43] The ideology of sportsmen regarding sustainable use of wildlife is protected by the wildlife management profession, which demonstrates that the NAM is in fact more a method of protecting special interests than a method for conserving wildlife.

While those affiliated with the NAM frequently state that hunting, fishing and trapping can only survive with broad support from the public, the only true and sole stakeholders that practitioners of wildlife conservation seem to be accountable to are sportsmen, not the public. Feldpausch-Parker et al. have observed that the NAM's tenets address as engaged citizenry merely consumptive users such as hunters and wildlife manager and scientists. However, facts are that there were 11.5 million hunters in 2016, a decline of 2.2 million from 2011,[44] with the number of an estimated 176,573 licensed trappers in 2014–15[45]—individuals who have purchased a trapping license— with that, hunters and trappers are a minuscule fraction of the general public in the United States. Yet, the NAM is defined by hunting and trapping, which makes its stakeholder group "at least 90% white and 90% percent male, and disproportionally rural."[46]

From the NAM supporters' perspective, this is a significant problem since the majority of Americans now live in urban areas[47] Other demographic shifts among the US human population predict that women will eventually outnumber men by far, and that white Caucasians will be the minority in the coming decade. While NAM promoters have actively been recruiting women and minorities into hunting and trapping, it is of significant concerns that while the NAM reflects a subset of the American population it is not widely applicable or interesting to the general public, which makes cooperation across society in support of conservation unlikely.

Nevertheless, the NAM is being promoted in a sustained manner that resembles a focused marketing effort. Another example is the National Hunting and Shooting Sports Action Plan (2016), which identifies its stakeholders as "Agencies, conservation and shooting sports organizations, and industry," and those entities already involved in the 3Rs – recruitment, retention and reactivation of hunters. To this end, the document identifies strategies and marketing efforts targeting families, women, and ethnic minorities to recruit them as hunters (and target range shooters).[48]

In the same vein, Feldpausch-Parker and her colleagues have expressed concern about the non-consumptive public and those unengaged in wildlife conservation apparently being asked to accept the NAM on its terms, because current efforts are focused on recruiting more hunters and preserving consumptive use of wild animals in the American and Canadian culture. Thus, they conclude that the NAM is a 'special interest' that competes directly and indirectly with other conservation ideals and organizations with non-consumptive or anti-hunting values.[49]

Non-hunting Shooters Are Now the Main Funders of P-R Monies

With the passage of the Federal Aid in Wildlife Restoration Act, known as Pittman-Robertson Act (P-R Act) in 1937, an exclusive stream of revenue, unaffected by Congress' whims, was made possible to fund state wildlife agencies and ensured their priority of serving hunter interests. Nie et al. (2017) explain that prior to 1937, "many states regularly diverted game license revenue to general governmental purposes, other than fish and wildlife management," however, Pittman-Robertson put a decisive end to this practice. Since then, the law provides states with federal money for wildlife restoration projects, natural resources management plans and hunter education,[50] but also "conditions federal funding on states using state game license revenue for wildlife management and conservation."[51] With the notable exception of Duda et al. (2017), publications on the NAM rarely acknowledge the substantial degree to which nonhunting shooters and archers contribute to Pittman-Robertson funds. Yet, as evidenced by a recent survey, a significant shift in funding sources for P-R funds has occurred; while the P-R Act was originally tied to 'game' hunting, a 2015 survey conducted by the National Shooting Sports Foundation showed that "shooters who do not hunt are becoming an increasingly larger percentage of the entire hunting and shooting population… and now represents an increasingly predominant segment."[52] Specifically, of the 51 million target range shooters, the segment of people surveyed, who participated only in sport shooting but not hunting, was the largest (44.2 percent) compared to those who only hunted (14.7 percent), or those who did both (41.1 percent). compared to those who only hunted (14.7 percent), or those who did both (41.1 percent). A clear gender gap emerged that showed the characteristics of individuals who hunt and target range shoot correlate with "male, 18 to 34 years old, and residing in a rural area in the midwest region of the country," while new target range shooters were primarily female, non-hunters, between 18 and 34 years old, and residing in urban areas.[53] Given these trends and in recognition of ongoing decline of hunters that would threaten the survival of the NAM if unchecked, the Modernizing the Pittman-Robertson Fund for Tomorrow's Needs Act (H.R. 877) was introduced in 2019. Among other benefits for hunters, this bill aims to ensure even more funding for already existing outreach programs as well as providing states with additional flexibility to implement a comprehensive agenda of nationwide R3 programs.[54]

NAM Overvalues Financial Contributions by Sportsmen, and Ignores Those from the Non-hunting public (Examples: PR-A and Public Lands)

As previously noted, the NAM promotes a "user-pay, user-benefit" concept, whereby sportsmen are required to purchase hunting, trapping and conservation licenses for the opportunity to hunt and trap wild animals. Funds derived from the sale of these licenses and through the Pittman-Robertson [P-R] Act) are commonly used as examples of the contributions made by hunters in support of conservation. Claims like these serve to influence the public into believing that hunters are conservationists, and to elevate their role as primary financiers of conservation. While it is true that these two funding sources comprise about 60 percent of state wildlife agencies' budget on average, there is a difference between what state wildlife agencies do—from administering license sales, securing access to hunting, fishing and trapping to stocking programs and controlling predators and more—and what constitutes conservation on the ground. Because a significant amount of this money pays for hunting-related opportunities, the conclusion that sportsmen pay for conservation is misleading. Secondary, more than 40 percent of state wildlife agencies' budgets come from non-hunting-related sources such as general funds, lottery receipts, speeding tickets, and taxes from general sales, outdoor recreation equipment and income tax check-offs, and other sources. There are multiple other ways the general public contributes to conservation, for example through membership fees and/or donations toward non-hunting organizations promoting wild animals and habitat conservation (e.g., the Wilderness Society, Western Watersheds Project); contributions by the public from the National Park Service and other financial contributions to supporting the public lands system and benefits to wild animals; many people are actively and rather selflessly engaged in the rehabilitation of wildlife, and contribute financially or otherwise to organizations that have a non-hunting and/or non-trapping policy on the land they preserve (e.g., the Wildlife Land Trusts managed by the Humane Society of the United States (HSUS), or the Raincoast Conservation Foundation in Canada.

With the exception of the National Wildlife Refuge System, funds derived from hunting-related activities have not contributed to the acquisition or management of federal public lands. An estimate by Mark Smith and Don Molde (2015) with Nevadans for Responsible Wildlife Management, shows that the non-hunting public contributed (through general tax revenues) about 95% of an annual $18.7 billion USA costs associated with federally managed public lands.[55] Sales from The Migratory Bird Hunting and Conservation Stamp Act of 1934 (commonly referred to the "Duck Stamp Act") show that hunters' fees have funded the purchase of about 1.1 million ha (3%) of the National Wildlife Refuge System (USFWS October 15, 2015), however the vast majority of lands in the National Wildlife Refuge System were acquired through means not related to hunting.[56]

In sum, the public is already actively engaged in non-hunting activities that provide benefits to conservation, such as rehabbing wild animals, birding, wildlife watching, outdoor photography, and hiking mechanisms for funding the acquisition and management of public lands, providing support for environmental legislation and conservation organizations that benefit wild animals. Yet, the NAM continues its biased portrayal and factually incomplete promotion of the "hunters-pay, public benefits" slogan, again implying that hunters are the main financiers of conservation and carry their weight while the non-hunting public does not.

An example provided by Thomas Serfass, Professor of Wildlife Ecology in Maryland and colleagues (2018) however, shows that instead, it is hunters who benefit hugely from the use of the public land system largely financed by the non-hunting public. The scientists explain, that "… although the acquisition and management of Bureau of Land Management and USFWS lands (39% and 14% of all federal lands, respectively) are not financed by hunting, about 95% of these lands are open to hunting."[57] The example wildlife refuges that are open to hunting and trapping has already been mentioned in Chap. 1.

Sportsmen: Portrayed as a Homogenous Group Against the Public

Another strategy by NAM supporters to protect their interests is to equate sportsmen as a homogenous group of conservationists, implying full, undivided and unquestioned support (financial and otherwise) for wildlife conservation. For example, the USFWS website states, "Hunters are among the most ardent conservationists around."[58] However, in their scrutiny of the NAM, Thomas Serfass and colleagues addressed this issue stating that this is yet another claim that has not undergone meaningful scrutiny and pointing to evidence suggesting the contrary. Clearly, a variety of sociological, psychological factors and personal preferences reflect different motivations to hunt, and with a certain number of sportsmen involved in game farming, fur trade and shooting reserves, trophy hunting, in support of killing wolves and other predators, while others are not. Given this diversity among sportsmen, Serfass et al. suggested a series of questions that would address this deficiency especially in light of the fact at hunters have never been given the choice of whether to pay or not. One of the questions suggested to ask, is: Do hunters disproportionally (in contrast to the non-hunting public) support environmental legislation that benefits wildlife conservation (e.g., the Land Water Conservation Fund, Endangered Species Act, and Clean Water Act), and vote disproportionally for politicians supporting this type of essential legislation and other non-hunting conservation initiatives? A 2012 survey suggested that traditional stakeholders will fight back in order to protect hunting and against changes of the current wildlife management regime towards a more conclusive and holistic, bio-diversity-focused state wildlife governance system.[59] This backlash of traditional sportsmen is a dominant cause for the current disasters of wolf slaughter in the Western states of the United States (to be discussed in Chap. 6). Also, considering the powerful interests, financial support, and lobbying work by the Conservation-Industrial Complex against the protection of wild animals, the likelihood of positive responses to these suggested inquiries is rather questionable.

NAM Is Not Law

In their publication, *Fish and Wildlife Management on Federal Lands: Debunking State Supremacy,* Montana-based Professor of Natural Resources Policy, Martin Nie and colleagues underscore the fact that the NAM is just a set of principles without any grounding in law or regulation. Yet, it is frequently used by state wildlife agencies to emphasize the importance of hunting, hunter-access, and the revenue from sportsmen licenses. This institutional bias "exacerbates the potential for intergovernmental conflict by displaying an institutional bias towards game species and hunters …" and b) to draw distinctions, rather than finding commonalities between state and federal priorities. Another main point of critique is that, despite the essential and increasing significance of federal lands in providing habitat for biodiversity, the NAM fails to include "a principle focused on habitat and the role played by federal lands in the conservation of wildlife—a failure that makes any account of conservation history, incomplete."[60] Given that wild animals are increasingly losing ground due to a range of human-caused factors leading to the fragmentation and loss of habitat, it is surprising, to say the least, that the creators and promoters of the NAM do not find it necessary to address the need for the protection of habitat and ecosystems.

NAM and Agency Capture

The term 'agency capture' describes the situation whereby governmental agencies are essentially controlled by those they are supposed to regulate, and then turned to their advantage.[61] Concerns related to agency capture include an undemocratic allocation of assets especially when special interests "capture the administrative agencies, or capture the constitutive process, defined as the rules governing wildlife agency decision-making." Reference In addition, selecting government officials for their sole commitment to sportsmen's interests increases the risk to be vulnerable to corruption, financial inducement, and political patronage. This in turn, leads to the degradation of assets and can undermine public trust in the asset's governance by the agency.

Indeed, a paper by Adrian Treves, Professor of environmental studies at the University of Wisconsin-Madison's Nelson Institute of Environmental Studies, and colleagues (2017) point to a significant problem related to NAM's conflation of hunting and the public trust doctrine (PTD). Specifically, the authors observe that the NAM's claim that its tenet #7 "Democracy of hunting" is commensurate with the public trust doctrine (PTD) has perpetuated and reinforced an illegitimate view of the wildlife trust in the USA and Canada.[62] Because of its relevancy of issues discussed in this book, particularly the NAM's hostility toward the animal rights movement, this section addresses it as succinctly as possible and relies heavily on the critical and comprehensive review by Treves et al.

We may recall from Chap. 3 that the public trust is mostly an unwritten[63] doctrine stemming from Roman law which originally had a rather narrow focus on the people's rights to common things, such as air, running water, the sea and the shores of the sea. Over the years the public trust has been expanded to include wildlife, parklands, beaches, and non-navigable waters and the public's interest in recreation, aesthetics, and ecological diversity. It is the principle that just, democratic governments must preserve environmental assets, including wildlife, and hold them in trust for current and future [human] generations. The PTD was envisioned to be flexible so that it can meet changing conditions and protect society from the injustice of undemocratic allocations and excessive use by a special interest group.

Legal scholars confirm a legitimate *'broad public interest view'* of the PTD supported by case law and other legal instruments. By contrast, the NAM promotes an extremely narrow view of the PTD that prioritizes consumptive, lethal use, developed and advocated by professionals in fish and wildlife management. The NAM's 'pro-hunting view' dictates hunting as the purpose of wildlife management, and hunters as privileged beneficiaries of the trust in wildlife. In other words, the historic interpretation of the Public Trust Doctrine and the NAM's more recent interpretation of the doctrine differ in three areas: 1) changes in societal values are comfortably accommodated in the PTD (broad public interest view) but not by the NAM; 2) the benefits of wildlife are shared equitably by all members of current and future generations regardless of their awareness, uses, value system, or fees they have paid (broad public interest view) versus the NAM's asymmetrical share with priority given to hunting, trapping, and angling; and 3) wildlife differs from other environmental assets because some interest groups seek eradication.

Given the core premise of the public trust doctrine, no public stakeholder group should benefit more than others from wildlife management. In reality however, sportsmen believe that because they 'pay to use wildlife,' they have an exclusive voice in the management of wild animals and to expect greater benefits from wildlife than those members of the public who do not pay. This claim has been long recognized as a major problem and has been described by certain wildlife agency officials as "… a potentially fatal, deeply rooted inconsistency between the rhetoric and reality in wildlife management in the United States."[64]

The pro-hunting consumptive view of the NAM counts social, legal, economic and ethical changes of society, in particular the animal rights movement, as a threat not only to the PTD itself but also to the entire institutional and legal framework of fish and wildlife management in North America. In addition, consumptive users claim that certain wildlife—predators such as wolves—are basically no longer a public asset but are to be seen as a (personal) liability for some (ranchers, sportsmen). Treves et al. (2017) have called the NAM's identification of the animal rights movement and a legally recognized concern (personal liability), and therefore the organizations espousing both, as existing threats to democratic institutions and the PTD itself. The authors have called this identification illegitimate and concluded that the pro-hunting view of the PTD and the NAM has failed to guide the trust managers in adopting a broad public interest view of the PTD. Treves and colleagues express a grave concern that, if adopted legally, the pro-hunting view of the PTD and the NAM would prevent future citizens from recognizing animal rights or personal liability concerns.[65] They also provided some suggestions for agency reform and public trust thinking in addressing the just allocation of environmental assets, ideas that are being advanced through a newly created organization called, "Wildlife For All," a national campaign to reform state wildlife management with a focus on ecology, democracy and compassion.

Conclusion

It should be of great concern to all of us when the preponderance of public servants working for governmental agencies, entities often seen as 'authority'—either directly or indirectly engage in the promotion of a self-interested agenda that is based on the recreational killing of wild animals, who are, allegedly, owned by all citizens, yet managed for a human self-interest group (some would call this a hijacking maneuver). It is of particular concern that government agencies mask the primary of meeting sportsmen's interests by accompanying, selective portrayals of benefits provided by hunters and hunting to wildlife conservation. Under such an approach to conservation, wild animals continue to be harmed and disenfranchised, while the general public is being kept outside the decision-making processes.

The North American Model for Wildlife Conservation empowers sportsmen, but not wild animals or their habitat. The NAM must be replaced with an approach to conservation that is inclusive to all levels of stakeholders—humans and non-humans. At a minimum, the NAM should be recognized as a stand-alone 'model' representing an ever-decreasing minority and should no longer be funded. Most certainly this approach should not be the paradigm of governmental agencies. Alternative approaches to conservations such as the one put forward by Compassionate Conservationists along with other ideas for non-lethal and co-existence concepts, which consider the question "what do wild animals need from us?" should be on the forefront of conservation, followed by creating broad public support for them.

Notes

1. Nelson et al., "Inadequate."
2. Ibid., 58–59.
3. Mahoney and Jackson, "Enshrining."
4. Posewitz, "Hunter's Ethics." 32.
5. Bixby, "Hunting Isn't Conservation."
6. Nelson et al., "Inadequate."
7. Laundré, John, "Are Hunters Conservationists?"
8. Wuerthner, "NAM and Wolves."
9. Jane Fonda, "Prairie Dogs."
10. Even the most cursory look online shows many videos on Youtube and in other places that show the bodies of prairie dogs literally exploding when shot.
11. Kuglin and Larson, "Pheasant Program."
12. Clark and Milloy, "North American Model," 340.
13. National Public Radio (NPR), "Extinction."
14. The Center for Biological Diversity's website notes that currently, 2244 species are listed as threatened or endangered under the Endangered Species Act, with 1618 of those in the United States. Of the listed species in the United States, 884 are plants; 307 are invertebrates, 163 are fishes, 95 are birds, 96 are mammals, 36 are reptiles, 35 are amphibians and two are fungi.

15. ScienceDaily, "Stark Warning."
16. Peterson and Nelson, "Problematic."
17. Serfass et al., "Empowerment."
18. Geist, "Creating Wealth," 286.
19. Nelson et al., "Inadequate."
20. Bixby, "Hunting Isn't Conservation."
21. Vucetich and Peterson, "Basic Principles," 5.
22. Trauger and Kennedy, "Pioneering Professionals," 32.
23. Chaney, "Family Pulls Award."

    RMEF has engaged in federal lawsuits supporting the removal of Rocky Mountain gray wolves from Endangered Species Act protection. In the spring of 2012, the organization also donated $50,000 to the Montana Department of Fish, Wildlife and Parks to fund contracts with federal Wildlife Services for more aggressive wolf-killing activities. This same article quoted RMEF's executive director, David Allen, having stated that "states should 'shoot wolves from the air and gas their dens' to control their populations."
24. Feldpausch-Parker et al., "Privileging."
25. Foran, *Subjugation*, 50.
26. Conservation Leaders for Tomorrow. "About."
27. Nie et al., "Debunking State Supremacy."
28. Ibid., 11.
29. Geist et al., "Defined," 179.
30. Geist, "Creating Wealth."
31. Nelson et al., "Inadequate."
32. Moag, "Rosalie Edge."

    Moag also writes that ECC's mailings like all Edge's advocacy, emphasized the need to protect complete ecosystems, rather than specific species. This was a dramatic shift from the orthodox conservation principles of the day, which tended to favor species with economic value or those that held the particular interest of powerful men. Birds of prey were particularly vulnerable to this way of thinking. At the urging of salmon fishermen, Alaskan territorial authorities in 1917 had offered a bounty of $1 or $2 for every bald eagle killed. Over the next 15 years, hunters shot 70,000 raptors out of the sky."
33. Ibid.
34. Fritzen, "About."
35. Peterson and Nelson, "Problematic," 3.

36. Organ, "Born," 24.
37. Peterson and Nelson, "Problematic," 3.
38. Feldpausch-Parker et al., "Privileging," 36.
39. Ibid., 36.
40. Ibid., 35–37.
41. Ibid., 37.
42. Ibid.
43. One agenda item on their 2020 "Hunting/Shooting Sports Participation Committee" meeting discussed a science-based survey publication by Mark Duda titled "How to talk to the public about hunting" recommending the use of "regulated" when addressing hunting, providing opportunities to connect non-hunters to hunting" and to "Engage animal welfare to combat animal rights" (Responsive Management Report Responsive Management 2015, p. 10).
44. US DOI, "Census."
45. AFWA, "Characteristics."
46. Peterson and Nelson, "Problematic," 4.
47. Parker et al. Pew Research Center, "Demographic."
48. Serfass et al., "Empowerment."
49. Feldpausch-Parker et al., "Privileging," 39.
50. For each of these categories, states are generally required to match 25 percent. For a breakdown of federal contributions to these categories and the shifting sources of P-R funding, see: https://responsivemanagement.com/wp-content/uploads/2018/10/The-Growth-of-Sport-Shooting-Participation-Edited.pdf
51. Nie et al., "Debunking," 12.
52. Duda et al., "Growth of Sport Shooting," 38–39.
53. Ibid., 40.
54. http://congressionalsportsmen.org/policies/federal/modernizing-the-pittman-robertson-fund

    An information sheet on the bill explains among other things that " ... female participation in hunting increased by 104% between 2001 and 2016, while female participation in target shooting increased by more than 81% during that same time as a result of increased R3 efforts."
55. Smith and Molde, "Wildlife Funding."
56. Serfass et al., "Empowerment."
57. Ibid., 106–107.
58. USFWS, "Hunters as Conservationists."

59. Manfredo et al., "Values."
60. Nie et al., "Debunking," 108.
61. Agency capture - Oxford Reference
62. Treves et al., "Predators and The Public Trust," 255.
63. The exception is the US state of Pennsylvania, which has included the PTD in its state's constitution.
64. Jacobsen et al., "Conservation Institution," 205–206.
65. Treves et al., "Predators and The Public Trust," 255.

5

The North American Model of Wildlife Conservation's Selective Use of Ethics to Support the Exploitation of Wild Animals

NAM and Anthropocentrism

The NAM has chosen a form of utilitarianism that is committed to an extremely narrow anthropocentrism that centers the maximizing of "undifferentiated desires,"[1] those of hunters, anglers, and trappers. Its utmost priority stands for personal and recreational utility of wild animals, a goal that is reflected in state and federal agencies' policies; the maximizing of opportunities for human citizens to kill wild animals via regulated hunting, fishing, and trapping seasons. Consequently, the NAM's centering of sportsmen's "desires" to hunt, fish, and trap animals demands a strong stance of anthropocentrism in order to attempt to justify its grand-scale exploitation of animals for the sportsmen's personal 'recreational use.'

Eileen Crist and colleague, Britain-based lecturer, conservation scholar and author Helen Kopnina caution that, when humans are centered, we need to ask who and what becomes displaced to the periphery. They explain that the long historic displacement of nonhuman animals can be categorized as ideational and material dislocations of nature and nonhuman animals to the fringes of physical landscapes and human

A. Heister, *Beyond the North American Model of Wildlife Conservation*, The Palgrave Macmillan Animal Ethics Series, https://doi.org/10.1007/978-3-031-14149-2_5

"mindscapes" alike.[2] The material dislocation of other-than-human nature is currently playing out globally. This "all-pervasive anthropocentricity," which is one of the "major cornerstones of modern social paradigm"[3] as well as the dominant ethic in wildlife conservation, is a root cause of the ecological crisis, human overpopulation, and extinctions of many non-human species. In the process, individual animals are being deprived of their life requisites causing them to suffer through stress, starvation and diminished security.[4] Two of Canada's most prominent conservation scientists, Paul Paquet and Chris Darimont have stated that "the same human activities driving the current extinction crisis are also causing suffering, fear, physical injury, psychological trauma, and disease in wild animals."[5]

With respect to the ideational aspect of exclusion and subsequent displacement of other animals on the human mindscape, Crist and Kopnina point to the long history during which humans have persistently inquired into the human identity by asking not what connects us but instead, how humans are *different* from other living beings. A long list of qualities exclusively possessed by humans that nonhumans lack has been proposed throughout this ongoing search for identity and entitlement, including language, morality, free will, and reason as perhaps the last bastion assumed to be setting humans apart from all others. They summarize,

> The dominant ideas of the Western tradition have primarily—and especially as a distilled and conditioning missive—not only exalted Man and his attributes but simultaneously portrayed nonhumans to be deficient by comparison. The ideational displacement has flown from the Differential Imperative's compulsion to story the assemblage of living beings on Earth as a hierarchical narrative.[6]

As Chap. 3 has shown, with the rise of the concept of sport hunting—enshrined as the NAM's foundation—came the fall (displacement) of wild animals into a legally-defined category of natural resources along with non-sentient natural entities such as trees, plants, soil, water and rocks. With the advent of the use of scientific management to study nature, the epithet of "renewable" was added when the wildlife profession realized that "wildlife is a natural resource that is renewable—a resource that, with wise management, can be perpetuated indefinitely for the enjoyment of present and future generations."[7] This self-serving view not only locates humans as outside and above the rest of nature, but assigned the role of sportsmen as *managers* of nature, in particular of other animals, while downgrading the status of wild animals to resources, on par with "other" natural entities, jointly called the "out-of-doors" thereby unyieldingly determining their status as inferior, and firmly cementing their value strictly for human use.

Thus, the theme that predominantly reverberates through the NAM's narrative, and characterizes its tenets and wildlife policy most distinctively, is its conviction that humans are subjects with moral standing, while wild animals have no such value other than being objects (resources), held as property in trust by state wildlife agencies for humans to use. Therefore, the desire of hunters, anglers, and trappers in "pursuing" wild animals for personal use and recreation, a non-vital human interest of some, dominates the NAM's concern, while the vital interests of the wild animal in surviving and not being harmed, are being pushed to the periphery of moral concerns. This is furthermore aided by its proponents' ethical allegiance to the whole (as opposed to systems' living parts)—the species or population, to the point that the individual animal is out of sight and entirely excluded. As Marti Kheel has pointed out, "[T]he sport hunters in the early conservation movement were intent on conserving species, *not individual* [emphasis added] animals in order to provide an arena in which they could continue their display of 'manly sport and adventure'"[8] Indeed, by replacing commercial hunting with recreational (sport) hunting, early sport hunter-conservationists were successful in retaining their main interest by *conserving* (sustainable use) wild animals, while *preserving* (protecting) these kill methods. NAM proponents often claim that they have replaced the earlier predatory ethos with an ethics of protection. However, Kheel has rejected this claim by pointing out that the distinction collapses when applied to individual animals. Instead, Kheel argues that,

> [T]he sportsman's code of early conservationists not only excluded other-than-human animals as individuals, it specifically sought to ensure the continuation of opportunities for violent confrontation with them. Thus, the

> new ethic of protection was designed to *perpetuate* predation. Instead of unlimited predation, the hunters of the early conservation movement substituted an ethos of controlled predation.[9]

Like Kheel, other authors have argued that human predation and protection are two aspects of the same worldview, both of which objectify other animals. Managed through violence and restrain (including setting seasons, establishing management units and bag limits, the request to follow regulations and competition rules, and so on), this type of management ensures a constant supply of animals to hunt and trap. The NAM's seven tenets reflect this extreme anthropocentric stance and instrumental attitude toward other animal species by declaring wild animals as mere human public/private property, as a natural resource (tenets 1, 2, 3, 5, and 6), or describe acceptable purposes for killing animals such as for "food, fur, self-defense or property protection" (tenet #4), and finally, the right of humans to kill animals for sport touted in the "Democracy of Hunting" (see Chap. 3, tenet #7).

Canadian wildlife biologist and internationally recognized authority on mammalian carnivores, Paul Paquet summarizes the ethical stance within wildlife conservation:

> This traditional management ethic favors an anthropocentric view that humans are an exceptional species and, aside from their utility for humans, other species are of little or no consequences in the large scheme of things. In traditional wildlife management, human domination over nature is the natural order. Nature is a commodity that is owned, and used by people, in pursuit of personal interests.[10]

According to Crist and Kopnina

> …a strong moral hierarchy deserves special attention because it is essential in defining action orientations: only by giving relative, and often absolute, value to the human over the nonhuman, and by withdrawing moral consideration for the other-than-human realm, can a reign of displacement, exploitation, killing and relentless use of the natural world (and its multitude of beings) become status quo.[11]

Indeed, the NAM's anthropocentric worldview and decision-making leads to hundreds of millions of individual wild animals being tortured, injured, and killed in the name of recreation, tradition, commerce and management—violence against and mass killing of wild animals—all of which run under the banner of conservation. This chapter now turns to two well-developed ethical theories in the Western tradition that have challenged anthropocentrism and addressed the moral exclusion of other animals and their treatment as a 'means to an end,' both of which are characteristic of the human (sportsman)-relationship with wild animals and promoted by the NAM.

Animal Liberation: Approach by Peter Singer and Tom Regan

Two ground-breaking books, *Animal Liberation* by preference utilitarian[12] Peter Singer and *The Case for Animal Rights* by rights theorist Tom Regan,[13] galvanized the animal advocacy movement in the mid-seventies and mid-eighties respectively when both philosophers challenged the anthropocentrism of Western tradition that grants moral standing exclusively to human beings and considers the rest of nature, including non-human animals, as sources of utility, or mere means for human ends. Singer and Regan provided formative influences on the evolution of the modern animal rights movement and a theoretical and practical path toward the liberation of nonhuman animals from human oppression. Armed with rational arguments presented by Singer and Regan, the animal liberation movement entered the field of ethical behavior and society by arguing that nonhuman animals have legitimate interests, and thereby legitimate ethical claims, regarding the aspects of human behavior ad society that affect them.

Still, while Singer and Regan share the goal of inclusion of other animals in the moral sphere, their theories are also markedly different in their approach. In broadest terms, Singer works within and extends utilitarian ethics to include a calculation of other animals' preferences, while Regan rejects utilitarianism entirely, arguing that nonhuman animals have an equal right to respect and that it is morally impermissible to utilize them as a resource. Singer and Reagan both follow an essentialist view of human and nonhuman beings in their proposal that possessing 'interests' is a single capacity that exists cross-species and a necessary condition for humans and nonhumans alike to be brought into the moral fold. Once assessed, they conclude that beings with interests have inherent value and therefore have a right to be treated with respect.

Central to Regan and Singer's theories is the concept of human "marginal cases." This argument observes that we tend to give equal moral status to marginal cases (rationally disabled humans, for example, those with senile dementia, severe mental impairment, or the comatose) while simultaneously denying equal moral respect to animals who may have cognitive capacities that are assumed to be comparable to these. In order to bring other animals under the umbrella of moral concern, Peter Singer and Tom Regan settle on the sole capacity of interests as the lowest common denominator shared by "normal" human adults and marginal humans. Regan argues that "mentally normal mammals," one year or older, are "subjects-of-a-life" with "preferences" (things that an individual desires, wants or dislikes and does not want) and "welfare" interests (things that would benefit an individual whether the individual is interested in them or not).[14] Subjects-of-a-life have an equal right to be treated respectfully according to the kind of value they have, and are owed this treatment equally; specifically, they are not to be treated as if they were mere "receptacles of valuable experiences."[15] Singer points to sentience, the ability to feel pain, joy, pleasure, and suffering, as the prerequisite for having interests.

Singer argues on the basis of equal consideration of interests and insists that species membership is morally irrelevant when pain and suffering are involved. Thus, human-inflicted suffering on animals who can experience pain is morally indefensible, and justifying these actions by arguing that animals are not human and therefore do not merit moral consideration is simply *speciesism*, "an irrational bias toward members of one's own species and against members of other species."[16] Speciesism, like racism, sexism, or genderism, uses morally irrelevant differences to justify treating nonhuman animals differently than humans, the core idea being that all forms of discrimination involve harming others on the basis of arbitrary and irrelevant characteristic (e.g., skin color, gender, or species).

Singer quotes the eighteenth-century utilitarian Jeremy Bentham to argue that the relevant criterion is sentience—the animals' ability to experience pain and pleasure. Bentham says, "The question is not, Can they reason? nor, Can they talk? But, Can they suffer?" The answer, of course, is a resounding "yes." Singer maintains that like interests ought to be treated in like manner and asserts that all sentient beings have an interest in avoiding painful experiences and acquiring pleasurable ones. According to Singer, possessing those interests entitles individuals due weight in a utilitarian calculation of both the animal and human pleasures and sufferings that result from exploitative practices, including vivisection, 'meat' eating, recreational hunting and trapping.

Regan, in his rejection of utilitarianism altogether, much more stringently argues that these practices violate an animal's right to respectful treatment, that is, we do not treat them in ways consistent with the recognition of their equal possession of inherent (intrinsic) value. Instead, we treat them as receptacles of instrumental value, lacking any value of their own. Finally, Regan distinguishes between moral *agents* (individuals capable of making rational, moral judgements, such as adult humans) and moral *patients,* (those incapable of making these judgements, such as children, or mentally incompetent adults, who nevertheless are entitled to be treated as ends). Regan argues that to arbitrarily exclude intelligent nonhuman animals from moral consideration as moral patients whose intrinsic value and rights must be recognized is, again, to be guilty of speciesism. Hence treating animals as ends in themselves is a moral duty, a matter of justice, not kindness.[17] Therefore, it is morally indefensible to treat animals as mere props for a human hunting or trapping experience. In summary, the traditional Animal Liberation theories put forward by Peter Singer and Tom Regan have shown that if one accepts human rights (considered inalienable) in the absence of a morally relevant distinction between other mammals and human beings, we cannot deny extending rights to a broader category of species without being guilty of speciesism.

Approach by Ecofeminism

While appreciative of Singer and Regan's contributions in placing the ethical discourse of the defense of other animals on the academic, social, and political maps, ecofeminists who are concerned about our treatment of other animals have nevertheless questioned certain premises inherent in both theories.[18] One particular shortcoming centers on the exclusively rational bias in their theories, which leaves out emotions and affections. As Singer and Regan both work within a justice framework, they present our relations with other animals as a matter of consistency and fair (equal) treatment of humans and nonhuman animals. Marti Kheel has critiqued the central approach in Regan's and Singer's theories of deriving moral status for animals from "marginally-rational cases" to show the limits of reason that both philosophers rely on to build their case for moral inclusion of animals. However, Kheel opines that the case for why we should accord rights to marginally rational human beings in the first place is never established. She concludes that the most important questions about moral relations to others cannot be answered through reliance on rationality alone. After all, rationality, including rules and principles, is in fact based on distinct and strong feelings and it is here, where we are much more inclined to accord moral status to other animals than deny it to the (marginal) humans.[19]

Also, as essentialists, the approaches by Singer and Regan are based on searching for "sameness" among the species, using humans as the moral yardstick against which all other species are inevitably measured. This again, leads to a ranking, whereby species that show the closest similarity with humans have a much better chance of having their interests considered by humans, while those unlike humans, including the non-sentient parts of nature (on which all life depends), are excluded with the consequence that those outside the circle will continue to be sacrificed for those who have made it in. As a result, this leaves in place the hierarchy of humans on top against which all other species are being ranked based on their similarities with what Singer and Regan have considered human capacities (actually traits usually associated with the human male and used by men to separate themselves from animals and women)—self-awareness, consciousness, theory of mind, memory and complex planning, communication, and intelligence.

Furthermore, as influential legal scholar and ecofeminist Catharine MacKinnon demonstrated in her reflection on the original strategy of the women's rights movement, the question, "Are they like us?" has been the wrong question for women, and it is the wrong question for animals, for neither of them will be met on their own terms as long as they have to show sameness as white male human beings in order to be recognized as full beings.[20] We need to consider these related questions: Why do women have to be like men to establish their equality, and why do animals have to measure up to human-defined standards and even resemble human beings in their physical and cognitive make-up in order to be free of the atrocities we inflict upon them?

Philosopher and poet Deborah Slicer has explained that searching for sameness rather than celebrating differences between human animals and other animals results in others being assimilated into the human sphere and, in the process, becoming honorary humans in different shapes and skins. And not only that, what is lost in the process is the acknowledgment and celebration of difference or "otherness," and with it the possibility of humble curiosity about other life forms, inquiry about their needs and interests and to appreciate and maintain *their* way of life. Furthermore, Slicer observes, "There is no reason why animals' differences, independence, indifference cannot be grounds for caring, for relationships characterized by such ethically significant attitudes as respect, gratitude, compassion, fellow or sisterly feeling, and wonder."[21]

Most ecofeminists who are concerned about the treatment of other animals agree with Peter Singer and Tom Regan in their striving to liberate other species from the oppressive grip of humanity, but they have nevertheless recognized that neither of the theories by Singer and Regan cut deeply enough into the anthropocentric (and androcentric) fabric of Western human society to effectively supplant a dualistic worldview, which, along with its power relations, maintain an all-pervasive domination by humans, and particularly by men over the rest of the natural realm. In recognition of gender-specific conceptions of morality, ecofeminists have developed the ethics of care as an alternative to the traditional rights approach.

Contributions from Ecofeminism

Drawing from insights of ecology, feminism and socialism, ecofeminists have long identified the interconnections between the oppression of disadvantaged humans, nature, and animals. American feminist and author Carol Adams specifies that in a patriarchal paradigm the representation of the human is de facto "white human maleness"[22] where the *other* is defined as everything that white human maleness negates: other races, gender, or species and nature in general, all of which are systematically used and abused. Inequality is maintained through a gender-specific use of value hierarchies, value dualism, and a logic of domination.[23] Underlying such a conceptual framework "is a set of basic beliefs, values, attitudes, and assumptions" which compose one's worldview, and some conceptual frameworks that are oppressive. An oppressive concept is one that "explains, 'justifies', and maintains relationships of unjustified domination and subordination."[24]

Value dualisms create either/or thinking that establish only two opposite ways of being, in which only one is valued. Value hierarchies create up/down thinking that allow for certain positions, ideas, and groups to be valued over others. Dualistic thinking is stressed in a patriarchal society, where all that matters is the maintenance and justification of male gender privileges and specific power distributions that benefit those already in power. Ecofeminists emphasize that at the core of Western patriarchal society are destructive practices rooted in a "logic of domination," which describes the most essential characteristic of an oppressive conceptual framework. According to feminist environmental philosopher Karen Warren, "…this system of thought assumes that superiority justifies subordination… and it provides a justification for keeping Downs down…" In a patriarchal system the Ups are those in power, usually socially empowered men, who will keep the Downs—vulnerable beings who need to be kept subdued and controlled, usually women and animals—down—. Without a logic of domination there would be just a description of differences and similarities.[25] "An important corollary of the ecofeminist position is that since the domination of woman and nature is conceptually twinned, whatever serves to oppress one acts similarly on the other. Likewise, whatever liberates nature will relieve the oppression of women as well."[26]

Carol Gilligan's landmark book *In a Different Voice* (1982) examined gender-specific differences in how women and men relate to the world. Gilligan's research identified a feminine conception as a morality of responsibility, contrary to a masculine morality of rights, which emphasizes separation rather than connection and focuses more on the autonomy of the individual than on her context and relationships. Thus, many ecofeminists ground animal defense and advocacy in an ethics of care, which was developed as an alternative to the rights approach[27] and emphasize emotions such as love, empathy, and compassion for human-animal connections.

This, however, should not be understood as an either/or approach (either rights or care) but rather as complementary. By paying close attention to the particular context of individual animals, the feminist care-ethics approach to animal ethics also includes a political analysis that exposes the reasons that cause animal suffering, abuse, and exploitation in the first place. Donavan and Adams have explained that the ethic-of-care approach is supported by certain core tenets. These include the awareness that "it is wrong to harm sentient creatures unless overriding good will result *for them*" and "it is wrong to kill such animals unless in immediate self-defense or in defense of those for whom one is personally responsible." Finally, we all "have a moral duty to oppose and expose those who are contributing to animal abuse."[28] Other ingredients of an ethic of care approach are the inclusion of the Animal Standpoint and the importance of compassion in our interaction and connection with other animals both of which will be discussed further in Chap. 8.

Ecofeminist Critique: NAM Perpetuates Separation, Dualism, and Objectification

In stark contrast to King's perception above, the NAM has imposed a strong hierarchy on nature, in particular with respect to wild animals. Kheel discusses several ideas that influenced the early 19th-century sport hunters: (1) the belief in a Darwinian view of nature as an arena of conflict and struggle; (2) the view of sport hunting as an evolutionary advance over subsistence hunting, as well as over the hunters from the lower classes; (3) the view that hunting helped to develop manly character, (4) the belief that rational rules and regulations were important tools for managing the natural world; (5) the idea that a detached approach of science was superior to that of "sentiment" as a mode of interacting with nature; (6) the belief in the moral value of self-restraint; (7) the idea that "species' have greater value than individual non-humans, and (8) the view of hunting as a patrilineal, cross-generational inheritance, to be passed from father to son.[29] She argues that sport hunters organized these ideas into a series of hierarchical oppositions, serving as the foundation for a common identity: nature/culture; animal/human; female/male; savage/civilized; work and play/sport, and unregulated predation/rational management (protection). These sets have greatly shaped not only the sportsmen's but also the movement of traditional conservation, and thus shaped society's relations to wild animals.

NAM's Narrative as a Framework of Oppression

Seemingly autonomous from the rest of society, the NAM's narrative focuses on a small group of white historic leaders such as Theodore Roosevelt, Gifford Pinchot, and Aldo Leopold, along with the occasional mention of one female, Rachel Carson, as a token woman (who was not a hunter but just too influential to be ignored). They highlight the important leadership played by early sportsmen's clubs, which consisted of men who "were decidedly patrician, representing the wealthy and elite."[30] And while these privileged men were busy figuring out how to retain sport hunting as their pastime, (and took it upon themselves to draft, enact, and enforce the first laws against market hunting of game), women at the turn of the 19th century were still struggling against ubiquitous oppression by white men and fighting to be recognized as a full people with a right to vote, while tens of thousands of native Americans, driven off their homelands onto marginal areas, were still reeling from the settlers' brutality and enduring the forced 'cultural assimilation into American civilization.[31] On the other hand, other white men put their privilege to good use. Henry Bergh fought against the cruel treatment of animals and founded the American Society for the Prevention of Cruelty to Animals in April 1866, three days after the first effective legislation against animal cruelty in the United States was passed into law by the New York State Legislature. Bergh also prompted the formation of the Massachusetts Society for the Prevention of Cruelty to Children in 1874.[32] Privileged women such as Rosalie Edge--a staunch opponent of hunters and trappers--were actively involved in the movements to advance the rights of women, nonhuman animals and wild animals (see Chap. 3), yet their efforts are entirely ignored and omitted in the NAM's narrative.

In response to the statement made by former president of The Wildlife Society (TWS), Bruce Leopold that, "the North American Model of Wildlife Conservation was not written at once like the Declaration of Independence,"[33] Josephine Donovan explains,

> [I]n the articulation of Locke and the framers of the U.S. Declaration of Independence and Constitution, not all humans were in fact considered sufficiently rational as to be held "persons" entitled to rights; only white, male property-holders were deemed adequately endowed to be included in the category of personhood. Indeed, much of the nineteenth-century women's rights movement was devoted to urging that women be considered persons under the Constitution. Here, as elsewhere in Western political theory, women and animals are cast together. This is also shown by Aristotle, who linked women and animals in the Nichomachean Ethics by excluding them from participation in the moral life and the century-long debate over whether women have souls "paralleled similar discussions over the moral status of animals.[34]

The gender-specific perspective of the NAM is, perhaps more than anywhere else, evident in the very story it is built upon—the man-the-hunter myth. This is one of the several origin stories that ecofeminists have identified as oppressive, as these narratives often serve to justify the oppression and exploitation of women and animals to varying degrees.

Although it is well known that there have been steps in developing civilization, the man-the-hunter narrative insists that the hunting and killing of animals by male hominids was crucial for the advancement of human society. NAM proponents often refer to hunting as an activity with a long history reaching back to Neolithic and Paleolithic times. Not to dispute the long history of hunting done by males, a selective and predominantly male-centric view of human evolution as being dependent upon man-as-hunter entirely ignores the contribution of the traditional role of women as gatherers, which, according to many anthropologists, is much more efficient and provides regular and consistent nutrition that sustains a group of humans. An accurate representation of humans' evolution would not only include the gathering of plants, roots, and other plant material, but also time periods when 'men were the hunted,' then scavenging flesh from animals who had died of natural causes or were killed by other animals was likely part of hominids' diet. Other concerning aspects of the man-the-hunter story are the presumptions that every being is naturally in conflict with one another; the establishment of men as the naturally aggressive and thus superior gender, and the overstatement of the role of hunting and the consumption of animal flesh (meat) in human evolution, all of which have been refuted by anthropologists' and ecofeminists' findings (for example, see Sussman 2006). Clearly, the exclusion of women —likely at least half of the historic citizenry—in the narrative favored by the NAM is informative in that it illuminates the NAM as having a colonial, male-centered approach with an instrumental attitude toward nature in general, and more specifically engendering an exploitative, violent, and destructive attitude toward wild animals.

NAM's Rationalism Framework

The NAM presents its view of the human relationship with the rest of nature, particularly other animals, as a nearly exclusively rational framework of seven principles, where emotions and affections are suspended in favor of principles, rules, and mechanistic, detached science. This stands in stark contrast with the often highly emotional testimonies by sportsmen trying to explain why they hunt (or trap). The Ecofeminist critique applies here for the same reasons the animal rights theories have been found to be insufficient. First, ecofeminists reject an inter-species morality grounded in abstract principles while leaving out emotions, affections, and virtue. A relationship with other living beings constructed this way is insufficient in that it does not realize that all abstract and rational tenets, in reality, are tied to emotions. Deborah Slicer explains that "[W]hat principles we articulate and ultimately choose to rely on are relevant to a very complex web of 'beliefs, feelings, modes of expression, circumstances and more, arranged in characteristic ways and often spread out over time.'"[35]

Second, an exclusive focus on rationalism at the expense of emotions and affections is a misrepresentation of human interrelations either with other humans or with nonhuman animals. Specifically, as Slicer explains, "this rationalist ideal fails to account for what motivates us in many of our personal relations, where love, or friendship, or the affections, for example, often are, and should be, a (or the) predominant motive."[36] Kheel has emphasized the limitation of relying on reasoning only by pointing out that those who have no regard for life cannot be persuaded by rational arguments. Rather, "those who *feel* their connection to all of nature to begin with, will take an interest in its continuation."[37]

NAM proponents often insist that sportsmen find connection to the natural world within nature. However, it seems that in the NAM's anthropocentric framework, "rationally" designed principles serve the purpose not to allow emotional bonds with living, feeling, individual animals, only dead ones. Here, as an expression of power—a human (a prisoner in isolated in cognition without connection to empathy) becoming God-like as the decision-maker of someone's life or death. Where the consciousness of the animal was necessary to measure human abilities against, this importance fades after the conquest and kill of the animal, and what emerges in its place, is that the animal body transitions into a materialistic possession, e.g., a trophy. This event floods the brain of the shooter or trapper with 'feel-good' chemicals/hormones for a 'job well done' on the inside while signaling his status to the outside (see SDO below). Feminists have also explained the links between hunting and predatory heterosexuality. Confirmed by countless anecdotes told or written by hunters expressing sentiments about their passion of and love for hunting, feminist Kelsey Brown observes, "Many hunters experience sexual arousal and entitlement to the bodies of those they can prey upon. In this sense, human women and animals face a similar relationship with people who embody this type of masculinity."[38]

Shane Mahoney, one of the most outspoken advocates of the NAM sheds light on this distorted connection with the dead clear when he describes that,

> [W]atching wildlife is only one-dimensional. The kill connects man in time with the prey (fish or wildlife), which includes the past as well as the present. A viewer only sees the creature in the present as it passes into and out of view. The hunter or angler, when killing the prey, participates in its past, knowing where the fish or game came from, and knows the life processes that bring that prey to the place of the kill. Hence, the hunter's connection with nature.[39]

Mahoney's conviction that the hunter is connected through killing an animal more likely implies an actual disconnect with the rest of nature, which is characteristic of the "masculinist paradigm" belief systems that are embedded in "the broader concepts of rationality, universality, and autonomy" without an emotional connection.[40] Here, Mahoney expresses the ubiquitous lack, or rather an absence, of a positive emotional connection (e.g., empathy) with the individual animal that is characteristic for the NAM.

Kheel opines that gender differences in violent, antisocial behavior, which manifests itself in the statistics reflecting a high percentage of males committing violent crimes, need to be addressed in moral thinking and that we cannot discuss moral conduct toward nature without acknowledging these same gender differences in violence toward nature. She points to object-relations theory that has emerged in feminist psychoanalysis for an explanation for the destructive male pathology underlying the ongoing violence against non-maleness. According to this concept, the early experience of boys and girls is to feel "oneness" with the mother figure.[41] Both will eventually come to their own identity separate from the mother; however, while the girl child remains feeling connected to the mother, the boy child comes to a negative identity which he establishes in opposition to everything that is female, the mother, women, and, by extension, nature. The socialization process then continues for the boy in his efforts to find his identity by not only transcending any emotional connection with the *others*—women, animals and nature—but by finding his identity diametrically opposed to all that is non-male.

Manhood, Killing Animals and Social Dominance Orientation

On his 2007 book, *Brutal: Manhood and the Exploitation of Animals*, a study on the concept of manhood and how it relates to hunting, animal experimentation, and animal sacrifice, feminist Brian Luke writes the following: "The construction of a manhood based on domination proceeds from an historically contingent presumption, namely that male identity must develop in opposition to female identity. If women generate life directly and positively, and men cannot be like women, then it follows that men's way of generating life must be indirect and negative, proceeding via the controlled destruction of others."[42] This results in a polarized labor division—women create life while men cause death"—in which animal sacrifice (e.g., hunting, vivisection, killing for 'meat') etc., is the way for men to establish paternity; they become generators of life through the ritual killing of an animal.[43] A growing body of literature is pointing to the importance of gender differences and highlighting the gender-specific approach in our relations with nature and other animals. Specifically, studies show that in general, women show higher levels of concern for the environment and animal suffering than men. Graça et al. (2018) provides further evidence that exploitative tendencies towards the more-than-human world is linked to higher levels of social dominance orientation[44] and lower levels of empathy in men compared to women. In turn, it is because women show less interest in group-based dominance and inequality and higher levels of empathy for other animals, that they show less support for animal exploitation and beliefs in human difference from and superiority over other animals. In short, higher levels of social dominance orientation and low levels of empathy result in support for animal exploitation and anthropocentrism (superiority/supremacy beliefs).[45]

Animal-Human Separation

NAM also follows a tradition of discontinuity—animal-human differencing—where the human self is defined as separate from the rest of nature, particularly from other animals. As recognized by ecofeminists, the underlying cause of the mistreatment of animals, as well as of the exploitation of women and the environment, is rooted in the post-medieval, Western, male psychology.[46] This is an argument that Carolyn Merchant made in her groundbreaking book, *The Death of Nature: Women, Ecology, and the Scientific Revolution* in which she warned about a male-configured worldview and science that constructs nature as a machine rather than a living organism, thereby sanctioning the domination of nature and women.[47]

According to Val Plumwood, the view of nature as sharply discontinuous or ontologically separate from the human sphere is the view of the human part of the self that is outside of nature. The maintenance of the sharp dichotomy and polarization is achieved by the rejection and denial of what connects human with animal. It follows that what is taken to define humans and what humans should be striving for is not to be found in "what is shared with nature (e.g., the body, sexuality, reproduction, emotionality, the senses, agency) but rather, what is thought to separate and distinguish them—especially reason and its offshoots."[48] Plumwood explains that, although humans have both biological and mental characteristics, it is the mental rather than the biological that has been taken to define the "authentic and full human… The human essence is often seen as lying in maximizing control over the natural sphere… and in qualities such as rationality, freedom and transcendence of the material sphere," which is, in short, the masculine ideal. Finally, Plumwood has emphasized that any critique of anthropocentrism (androcentrism) must include the connection of both to rationalism, because it is rationalism that grounds "inferiorization of both women and nature." The resolution is, then, to teach males, in particular young boys, to embrace those aspects usually rejected instead of insisting on differences and separation.

The Legacy of a White Hetero-Patriarchal Worldview

Defining humanity as separate from and in opposition to the rest of nature has a long history. The traditional philosophical argument and common view that animals can be exploited because they have no consciousness stretches back to Aristotle's idea of the Great Chain of Being, which ranked animals below man because they were thought of as unable to reason. This concept hierarchically orders matter and life, beginning with God on top, and then progresses downward through angels, men, animals, plants, and finally rocks.[49] According to this hierarchical staircase-like image, "man" is caught in a moral struggle between spirit and flesh where the former brings him closer to God, the ultimate spirit, and desire for the latter moves him away from God, resulting in the loathing of flesh associated with women, animals, and nature. Kheel has commented that:

> [T]he assumption that rationality and autonomy are unique to humans has been reinforced in the Jewish and Christian notion that only "man" is made in the image of God. As a consequence, only men (and by extension, women) have been viewed as "person" who are unique and irreplaceable creations worthy of dignity.[50]

The thinking of French philosopher and scientist René Descartes (1569–1650) has been particularly detrimental to our moral relations with other animals. Descartes believed that only humans had a body *and* an immortal mind, whereas other animals have only a "dumb" body that was lacking a mind, therefore, the animal is not an embodied soul. Descartes, as a dualist (mind-body split), was convinced that only humans had a body and an immaterial mind; bodies are devoid of thoughts and lack consciousness similar to trees or a dog's body; and a body that lacks a mind cannot feel pain.[51] Descartes conducted atrociously painful experiments on animals while claiming that an animal is nothing more than a machine. This Cartesian view of the world is what "grounds the common law doctrine of animals as property" (things).[52] Descartes' dreadful cruelty was, he insisted, a triumph of the Age of Reason, proving that animals had no capacity for thought or feeling which, in turn, reinforced the dominance of rational man. Over the last 400 years, most of our worldview has slowly caught up with the facts and we now know that Descartes was dead wrong about nonhuman animals and their allegedly lacking sentience and cognition while humans unquestionably (in his mind) possess these capacities.

On the Ontologizing of Animals

Though there are manifold existing critiques of Cartesian ontology, and human science and philosophy have evolved to accept animals as beings capable of cognition and sentience, the prevailing ideology in much of human culture ontologizes non-human animals as a resource to be used: whether as stand-ins for research methods we don't want to subject humans to, as living targets for recreational shooting or trapping, or as consumable food. We compartmentalize animals into various categories according to the usage we have for them; thus some may be labeled *research animals*, some are *food animals*, while others are *game* animals or *furbearers*. Carol Adams, feminist, animal rights advocate and author of *The Pornography of Meat*, among other publications, states, "[O]ntology recapitulates ideology." In other words, ideology creates what appears to be ontological: just as women are categorized as sex objects (the fur industry also makes use of exposing half-naked women's bodies wrapped in animal pelts), animals are categorized as *carriers of meat* or *bearers of fur*. This ontology is socially constructed, for there is nothing about animals in and of themselves that would necessitate their lives and destiny as uses toward solely human purpose.[53]

Adams explains that society uses language to make the subject disappear and hide the fact that there is a human perpetrator of violence in the process. Adams says, "When a living being is turned into 'meat,' language is structured to conceal: 'Someone kills animals so that I can eat their corpses as meat,' becomes 'animals are killed to be eaten as meat,' then animals are 'meat,' and finally, 'meat animals,' thus 'meat.'"[54] A similar process is used by the trapping and hunting community within the NAM where trapping and killing sentient wild animals for their fur becomes "wild fur trapping," and, wild animals are reduced to "wild protein." Animals who are being trapped as furbearers and hunted as "game" become a "crop" to be "harvested." The action of killing wild animals is referred to as "taking" and trapping and hunting are disguised as "tools" for "managing" wildlife populations. In our consumer-obsessed society, "pleasurable consumption of consumable beings is the dominant perspective of our culture. It is what subjects do to objects, what some*one* does with a something."[55] According to Adams, the most efficient way to ensure that humans do not care about the lives of animals is to transform living and feeling nonhuman subjects into nonhuman objects. When subjects are turned into objects, we suspend our emotions and we no longer care. The common view here is that animals are not like us (humans)—they are just animals. In other words, they are inferior and as such, we believe that neither their lives nor their deaths matter to them—that's what they are here for. Under such othering, humans are not required to feel empathy for things when we merely consume a food product—it's just food. Humans are then conveniently allowed to forget that no pig dies as pork, no elk dies as game, and no bobcat dies as fur. As Adams so poignantly notes, animals die as individual beings, and how they die "matters a great deal to the one who is dying."[56]

How the NAM Ontologizes Animals as Natural Renewable Resources

By denying the individual animal in its considerations, the NAM reflects Theodore Roosevelt's doctrine of consolidating all aspects of nature, sentient beings and non-sentient entities, into one whole—the "outdoors"—thereby lumping wild animals together and treating them as non-sentient things—as "renewable, natural resources," the same way as (non-sentient) features of the land such as trees, rocks, and waterways. The NAM's mechanistic worldview and instrumental attitude comes clearly into focus, where wild animals are considered things or non-feeling objects that can be used. According to this worldview, individual animals matter only as means to an end to meet the interests of certain, individual humans. It's not that the living and feeling animal is valued but that the human experience of having hunted or having trapped has value, and where maximum utility is the valued goal. A frequently used quote summarizes this truly callous sportsmen's arrogance, which begins with "To the sportsman, the death of the game is not what interests him .... One does not hunt in order to kill: on the contrary, one kills in order to have hunted."[57]

However, as Peter Singer and Tom Regan have already convincingly argued, other animals are sentient beings who must be included in any actions that affect them. Singer has explained that other species, particularly mammals and birds, display external signs of pain behavior, including "writhing, facial contortions, moaning, yelping or other forms of calling, attempts to avoid the source of pain, appearance of fear at the prospect of repetition."[58] He further points out that mammals have similar nervous systems, which correspond physiologically in ways humans do when we feel pain, ranging from an "initial rise in blood pressure, dilated pupils, perspiration, an increased pulse rate, and if the stimulus continues, a fall in blood pressure."[59] The capacity of feeling pain enhances a species' prospects of survival—dangerous or injurious situations can be avoided—and did not arise *de novo* in humans but developed from species already capable of suffering both from physical pain and from fear, anxiety, and stress.[60] Sentience obviously gives an animal an advantage in survival; thus, the NAM's focus on human interests that outweigh animal's sentience is morally wrong and the NAM's classification of other animals as non-sentient natural renewable resources is false.

Wild Animals as Human Property

The NAM views other animals as human property. The NAM has set up a legal system that has cast an all-encompassing net over wild beings who are self-willed and outside the legal realm in order to ensure humans' rights to use wild animals as common-turned-into-private property. For example, the Montana Furbearer Trapping Regulations reflect this biased constellation which protects a captured or dead animal in a trap as private human property while they address the interference with a trapped animal as private property, by stating, "[D]isturbing traps or trapped animals—"A person may not destroy, disturb, or remove any trap or snare belonging to another person or remove wildlife from a trap or snare belonging to another person without permission of the owner of the trap or snare ..."[61] In practical terms, this means that a wild animal, who at the time of getting caught has already transitioned into the trapper's private property, and already suffered in a trap or snare without any legal protection, is fully at the mercy of a trapper, who can take cruel behavior to its extreme by torturing a trapped animal at will before killing her by either stomping, strangling, or beating to death, shooting, poisoning, or drowning. As attorney and animal liberation advocate Gary Francione (2000) comments, once an animal's status is determined as property, what is being balanced are the interests of property owners against the interest of their animal *property*. The 'property-owner' constellation was evident during the era of slavery where slaves' lives were deemed irrelevant and cruel abuse and killing of these human beings was rampant. Thus, the legal status of property —and its accompanying view of moral inferiority—ensures that nonhuman animals' vital interests such as survival will never be given the same weight as even the most trivial interests of humans, such as hunting or trapping for recreation or commerce. In practical terms, this means that a wild animal, who at the time of getting caught has already transitioned into the trapper's private property, and already suffered in a trap or snare without any legal protection, is fully at the mercy of a trapper, who can take cruel behavior to its extreme by torturing a trapped animal at will before killing her by either stomping, strangling, or beating to death, shooting, poisoning, or drowning *his* trapped victim. Again, who is monitoring what sportsmen do the wild animals?

The NAM's View of Animals as a Population and Species, Not as Individuals

The NAM is not concerned about the individual animal but rather directs its moral allegiance to, as Marti Kheel describes it "the whole"—the animal species, and population. In this sense, the individual animal has no moral worth; she or he is just a functioning member of a greater system and it is the system itself which constitutes the NAM's locus of moral concern. In *Acts of Objectification, Ecofeminism, and the Ecological Narrative*, feminist Chaone Mallory notes critically: "The individual is only valuable insofar as it carries the genetic coding to perpetuate the species, which in turn is by evolution adapted to its surroundings and helps to perpetuate the healthy functioning of the ecosystem."[62] The impartial identification with everything can, in practice, care for nothing, which in turn, allows the sacrifice of millions of animals killed for sport, recreation, commerce, trophies, personal entertainment, and science. Particular emotions—for example, kinship—are suspended and suppressed in order to be able to kill.

Orienting the focus of concern in wildlife management and conservation on exclusively abstract constructs, such as species or population, ensures that the individual animal vanishes and can therefore be sacrificed for "the greater good" of the whole and for the experience of the individual human sportsman. As Kheel notes, "[T]he message of course, is that other-than-human animals are interchangeable or replaceable, whereas human beings are unique and inviolable."[63] The view of animals as "aggregate categories," or so-called mass terms such as *wildlife* (which encompasses species as varied and different as pine martens, great horned owls, river otters, ladybugs, and grizzly bears), while referring to large numbers of individual animals with the use of the singular, for example, *elk, wolf* or *bobcat*, all serves to make the individual animal invisible and to literally absorb the individual animal into the aggregate, who cannot hold opinions or sentiments as a group, since they do not have the ability to advocate as such.

Kheel has also noted that, "Just as men under patriarchal society view women as their antithesis in the quest for masculine self-identity, so too humans have often viewed animals as a foil for the establishment of human identity."[64] The joint oppression of women and other-than-human animals run parallel. This claim is supported by Adams, who has said it is "a cycle of objectification, fragmentation, and consumption... Consumption is the fulfillment of oppression, the annihilation of will, of separate identity."[65] The NAM's inherent dualism in denying animals individuality points, perhaps, to a more subtle problem—through sanctioning the killing of animals for the experience of the individual human being, it denies that specificity occurs not only between, but among the species. It is here that the NAM is guilty of employing a biologically unsound double-standard where one set of principles applies to nature and animals —the belief that only animal species count and no differences exists between individuals within animal groups—while another set of principles applies to humans—the belief that individuals count and that differences in human groups exist. Here, the NAM's blatant human/nature dualism, which privileges humans, comes to the forefront, and the non-human individual and their inherent subjectivity is declared irrelevant.

In stark contrast, the feminist ethics of care recognizes the diversity of nonhuman animals—one size does not fit all—with each having a particular history, and standpoint. This ethics is both biologically defensible and ethically consistent, and states that we need to pay attention to particularities in any decisions that affect them. The feminist ethic of care regards animals as individuals bestowed with feelings, who communicate those feelings, and to whom humans have moral obligations.[66] In the same vein, Kheel emphasizes that moral worth can exist *both* in the individual parts *and* in the whole of which they are part.[67] The point Kheel is making is that the sole reliance on reason forces competition and hierarchical thinking. Instead, if we allow positive feelings such as empathy and compassion in our interactions with nature, we can overcome these dualisms, and opposite positions "dissolve into different points on a circle."[68] Kheel's notion resonates with wildlife ecologist John Vucetich and environmental ethicist Michael Nelson, who have stressed the crucial role that empathy can play in dissolving the dualism between concern for the individual animal (usually the concern of animal rights advocates) and concerns for animal populations (generally expressed by conservationists) by acknowledging that a) the border between ecological communities and individual organisms is blurry, and b) "'respect for living things' is the root connecting concern for both," and that empathy can transcend them.[69]

Conclusion

An analysis of the ethics of the NM through the lenses of animal rights and ecofeminism exposes the model's blatant exclusion of non-human animals' interests and needs. Under such scrutiny, the NAM reveals itself as a bastion of outdated ethical bias and social domination based on human supremacy views that lead to exploitative, destructive, and violent behavior towards the rest of nature. Such analysis, in addition to advances in science and consensus in animal ethics call into question the practices the NAM perpetuates, and points toward a need to value and treat other animals as individual beings—subjects—with equal, species-specific rights, interests and needs. As it stands, the NAM representing the federal and state government's paradigm does not seek to increase our society's sense of connection and empathy—instead it is contributing to an existing empathy deficit.[70] Finally, in none of its tenets does the NAM explain why society should accept the undifferentiated 'desires' of sportsmen to kill wild animals for frivolous reasons over the right to life of the individual nonhuman animals.

Human society, individually and collectively, is accountable to the ethical claims of those who lay outside of it; in this case, the claims of nonhuman animals (as presented by empathetic humans). To assert these claims, both Singer and Regan have based their positions on revised traditional ethical models. By revamping these traditional ethical frameworks to include equal, direct consideration of the interests of nonhuman animals, they employ moral extensionism, a methodology for broadening the ethical arena by expanding the scope of tried and true ethical frameworks. Contributions made by ecofeminists have shown the crucial need for positive and connecting emotions such as love, empathy, and compassion in our relations with nonhuman animals (and nature in general). A closer examination of the NAM identifies it as an outdated paradigm that, at its core, is not substantiated through evolutionary theory either. It is this dangerous paradigm that has been driving the national and global extinction of wild animals and the destruction of their habitat and ecosystems and ultimately renders the NAM's ethics as inadequate in curbing society's ongoing destructive behavior toward the more-than-human world.

Notes

1. Van de Pitte, "Moral Basis."

   She also notes that "What is called for is, in addition to a fully understood accounting of the impacts of wildlife populations on our lives, an accounting of the trade-offs we are willing to make between pragmatically based desires and our insights into the kind of people we want to be, the lifestyle most conducive to self-respect, and the kind of community (denaturalized and hard technologized versus natural and soft technologized, etc.) we would prefer if we had the option. It is the complexity of determining what is worthy of desire, rather than surveying what people in fact desire, that leads to the construction of coarse utilitarian moral templates (wedded to classical economics) and thence to coarse political correlates" (pp. 264–265).
2. Crist and Kopnina, "Unsettling Anthropocentrism," 388.
3. Paquet and Darimont, "Same Coin?" 184.
4. Bekoff, Minding Animals.
5. Paquet and Darimont, "Same Coin?" 178.
6. Crist and Kopnina, "Unsettling Anthropocentrism," 388.
7. Bolen and Robinson, *Wildlife Ecology*, 3.
8. Kheel, *Nature Ethics*, 11–12.
9. Ibid., 82.
10. Paquet, "Crossroads," 1.
11. Crist and Kopnina, "Unsettling Anthropocentrism," 393.
12. The moral theory of utilitarianism states that what is right in a certain situation is dependent on the consequences of our actions and that we should choose the action which maximizes what is considered good for the largest number of those affected (Francione 2000). To a classical utilitarian like Jeremy Bentham (the founder of utilitarianism) happiness is the ultimate good (Francione 2000, pp. 131, 135). Whereas to a preference utilitarian like Peter Singer the satisfaction of preferences is the ultimate good (Francione 2000, p. 135). Singer points out that sentience is the essential characteristic for eligibility into the moral community. He illustrates this with Jeremy Bentham's famous testimony of animal's rightful place in the moral community: "... A full grown horse or dog is beyond comparison a more rational, as well as a more conversable animal, than an infant of a day, or a week, or even a month, old. But

suppose they were otherwise, what would avail? The question is not, Can they reason? Nor can they talk? but, Can they suffer?"

13. Although he is a follower of Kant, Regan breaks with the theory of Kant, who argued that "animals… are there merely as a means to an end. That end is man" (as explained in Donovan 1993, p. 170).
14. Regan (1983), *The Case*, 87–88.
15. Ibid., 277.
16. Singer (1977), *Animal Liberation*, 7.
17. Regan, *The Case,* 151.
18. For a more comprehensive critique of Singer and Regan from an eco-feminist perspective, see Deborah Slicer (1991, pp. 108–124).
19. Kheel, "Liberation."
20. MacKinnon, "Mice and Men."
21. Slicer, "Your Daughter," 112.
22. Adams (1993), "Feminist Traffic," 203.
23. Warren, "The Power and Promise."
24. Ibid., 127.
25. Warren, "The Power and Promise."
26. Mallory, "Acts of Objectification," 68.
27. Adams and Gruen, *Ecofeminism.*
28. Donavan and Adams, "Introduction," 4.
29. Kheel, *Nature Ethics,* 65–66.
30. Organ et al., "Fair Chase," 529.
31. World Heritage Encyclopedia, "Cultural Assimilation."
32. Faunalytics, "Bergh."
33. Leopold B., "Understanding," 12.
34. Cited in Gaard (1993, p. 169).
35. Slicer, (citing Walker 1989, p. 18).
36. Slicer, "Daughter," 115.
37. Kheel, "Liberation," 48.
38. Brown, "Feminist Analysis."
39. Elliot, "Mahoney."
40. Kheel, *Nature Ethics*, 3.
41. Ibid, 45.
42. Luke, *Brutal,* 23.
43. According to Luke (2007), cross-cultural studies across age groups of attitudes of Americans toward animals have consistently shown a significant gender gap where men generally support animal exploitation and oppose animal liberation, whereas women oppose the former and sup-

port the latter. Luke concludes that "… the struggle for animal liberation is also a struggle against a manhood defined by sexism" (p 23).

44. Graça et al. (2018) provides an overview of past findings that show that "women display greater levels of environmentalism partially because of their higher levels of empathy and lower levels of SDO than men … women tend to be more empathetic than men … women express lower desire and support for group-based dominance and inequality … that SDO and empathy are intrinsically linked to each other …. That SDO has been consistently linked to human-based hierarchical views towards the natural environment, as well as support for the exploitation of animals in favour of human interests and human supremacy beliefs" (p 66).
45. Graça et al., Why Women Are Less Likely."
46. Donovan, "Introduction," 65.
47. Merchant, "Death of Nature," xviii.
48. Plumwood, "Nature, Self and Gender," 10–11.
49. Nee, "Great Chain."
50. Kheel, "Liberation of Nature," 6.
51. Regan, *The Case,* 21–22.
52. Donovan, "Introduction," 238.
53. Adams, "Feminist Traffic."
54. Ibid, 202.
55. Adams, "Feminist Traffic," 13.
56. Adams, "War on Compassion," In: Feminist Care Tradition, 26.
57. Orion, "What Does it Mean?"
58. Singer, Animal Liberation, 11.
59. Ibid.
60. Rollin, *Unheeded Cry.*
61. Montana FWP, "Furbearer," 3.
62. Mallory, "Objectification," 69.
63. Kheel, *Nature Ethics*, 7.
64. Ibid, 172.
65. Adams, Sexual Politics of Meat, 47.
66. Donovan and Adams, "Introduction."
67. Kheel, "Liberation," 44.
68. Ibid.
69. Vucetich and Nelson, "In Review," 5.
70. Hendershott, "Empathy Deficit."

6

NAM's Science and Impacts of Policies in Pacific and Mountain West Regions

Aldo Leopold's Long Shadow

Although traditional wildlife management has progressed over the past 50 years toward integrating ecology and the whole biotic community, the NAM reflects the thinking and beliefs of the early 20th century. In his later years, and through his eloquent essays published in *A Sand County Almanac and Sketches Here and There* (1949), Leopold was instrumental in transforming the belief system of the conservation movement through his proposal of managing the land holistically by integrating science and ethics with empathy for the land. This approach no longer saw humans in the role of conqueror but rather as integral members of the biotic community and is encapsulated in his well-known quote: "A thing is right when it tends to preserve the integrity, stability, and beauty of the biotic community. It is wrong when it tends otherwise."[1] Despite being such a visionary, Leopold's ideas about wild animals never advanced beyond human supremacy thinking. Instead, as Foran writes, "Leopold reached into rationalization to justify the inclusion of hunting in his land ethic," and held on to beliefs about the need to control and manage the components of nature, ideas that continue to underpin contemporary policies directing wildlife management.[2]

A. Heister, *Beyond the North American Model of Wildlife Conservation*, The Palgrave Macmillan Animal Ethics Series, https://doi.org/10.1007/978-3-031-14149-2_6

The first American Game Policy was drafted by a national committee chaired by Aldo Leopold and presented at the American Game Conference in 1930. It began by stating, "[D]emand for hunting is outstripping supply. If hunting as recreation is to continue, game production must be increased." The policy draft emphasized the importance of hunters starting to pay "for their shooting" while recognizing the important role of landowners investing in land thereby producing game.[3] Shortly after that, Leopold's 1933 American Game Policy was published, which provided the foundation for the development of the NAM and for the state government management system, the inclusion of sportsmen and ranchers as the stakeholders, and an exclusive funding system that rewards the paying sportsman with the opportunity to kill wild animals.

Having reduced wildlife to 'crops to be harvested for recreational use,' Leopold's agricultural view paved the way for theoretical grounding of the NAM's science—its harvest theory with its objective to achieve a maximum sustainable yield (MSY)—the mathematical idea of predicting the maximum of surplus production can be removed while protecting some of the natural capital in perpetuity. This requires the manipulation of animal populations and species, as well as their habitat, for traditional wildlife management to generate a surplus of animals who then can be shot or trapped on a 'sustained' level.

NAM: Main Concepts

Contemporary wildlife management is frequently referred to as "the application of ecological knowledge in ways that seek a balance between the needs of wildlife and the needs of people. With the realization that the profession depends on "effective animal population and habitat management" came a focus on quantitative approaches to ensure "accurate numerical analyses of census data" to inform management decisions.[4] Also in this context, science is used to set hunting and trapping seasons, bag limits and quota of animals that can be shot or trapped per management unit, which in turn inform science.

Until the 1960s, wildlife management was primarily understood as game management. Today, the term wildlife is understood as encompassing "the practical ecology of all vertebrates and their plant and animal associates."[5] Vertebrates only comprise 3% of the approximately 1.5 million animal species that have been described, and terrestrial vertebrates, in particular birds and mammals—the focus of harvest management—make up less than 2% of all species.[6]

A wildlife *population* is usually understood as a collection of individuals of a species in a defined area that may or may not breed with other groups of that species in other places.[7] The term *management* clearly implies human manipulation of habitat and wild animal populations according to specified objectives. The common understanding of wildlife management includes maintaining certain wildlife species levels for the purpose of harvest, for the control of excess numbers of nuisance species, applied outcomes of population biology, ranging from measuring and interpreting trends, to setting harvest limits, to evaluating viability of endangered species and "determining the effects of predation in prey populations."[8] In wildlife management, sport hunting of birds (waterfowl) and antelope, deer, elk, moose and other game animal species, is viewed as central in informing conservation science. For example, the Lincoln-Petersen index, a mark-recapture estimator is used to estimate waterfowl populations.[9] Accordingly, birds are live-caught and banded prior to the hunting season and hunters are asked to report all shot birds that are banded in order to help with setting of seasonal bags and other information. In addition to this type of waterfowl monitoring, other scientific advancements include the estimation of survival rates (today specialized computer software is used), understanding density dependence, balancing predators and prey populations, as well as insights into survival variation, adaptive management, marking techniques and habitat enhancement.[10]

In traditional wildlife management, hunting and trapping are viewed as useful tools to inform science on predator-prey relationships by manipulating predator populations. Termed *removal* and *controlled harvest*, populations of predators such as wolves, pumas, or coyotes are frequently and to varying degrees "reduced" (killed, often extensively) in order to gain insights in the complexities of the relationships between predators and prey animals, since harvest by humans and predation are viewed analogously in many ways for their impacts on prey animals (Mills 2007).

Usually, the prey animals are also the game animals (trophy) hunters are interested in. Finally, the *adaptive harvest management* (AHM) approach, which relies on collaboration between wildlife managers and scientists in making management decisions and adapting these to changing circumstances "represents the most modern theory of harvest management to date."[11] As wildlife professionals frequently point out, a variety of techniques to catch animals, such as drop nets, Clover traps, darting and capturing via helicopter and others, have been developed from the study of abundant game populations. Funding provided by sportsmen is often also used to justify habitat manipulations (so-called enrichment), conducted by state wildlife managers to increase the number of animals available for sportsmen's guns, arrows, traps, and snares.

Wildlife harvest management includes estimating population characteristics, vital rates (including abundance, density, survival, reproduction and sex ratio), emigration and immigration can be represented by the BIDE equation: The abundance or size of a population (N) at time t+1 equals the abundance the previous time step, *t,* plus the number of animals that arrive due to birth (B) or immigration (I), and minus those dying (D) or emigrating (E).[12]

The harvest theory is literally at the NAM's foundation of hunting and trapping, and thus makes up the body of the wildlife management literature. The general theory of 'harvesting' animals is based on the following premise:

> Without harvest, growth and recruitment of the population are balanced by natural mortality; therefore the average growth rate of a population at carrying capacity is zero … Hunting reduces the number of animals in a population, but the loss also increases the growth rate. The increase in the population's growth is the consequence of higher birth rates and lower death rates, which result from decreased competition for food and other resources (i.e., with fewer animals, there are more resources per individual). Consequently, the accelerated growth rate provides a surplus of animals beyond the number required for replacing the losses—a surplus that may be harvested by hunters or other predators.[13]

Managing animal populations for sustainable harvest and long-term sustainable yield involves the key concepts of population stability and growth modelling the conditions under which it occurs. It also requires population manipulation in order to produce a maximum sustainable yield (MSY), which has been defined as "the greatest harvest that can be taken from a self-regenerating stock of animals year after year while still maintaining a constant average size of the stock."[14] It is assumed that a Maximum Sustainable Yield (MSY) can be taken continuously on an annual basis when a population is kept at half of its biological carrying capacity (K).

The formula is $H = Kr/4$ where maximum sustained harvest (H) is one-fourth the maximum population (K) times the intrinsic growth rate. In harvested animal populations that grow logistically, the MSY—the maximum number of animals that can be harvested—is possible at 0.5K[15] (this means that, while the rate of increase slows as the population reaches its carrying capacity, the rate of animal 'production' is greatest—the maximum—at half of its carrying capacity). In recognition of its inherent simplicity of complex processes, scientists have cautioned that logistic growth is just one way to describe how negative density dependence manifests itself in wild populations, and both positive and negative density dependence may be very common at different times and population sizes.[16]

Critique: Removal and Controlled Harvest

Termed *removal* and *controlled harvest*, populations of predators such as wolves, mountain lions, or coyotes are frequently and to varying degrees "reduced" (killed, sometimes extensively) in order to gain insights into the complexities of the relationships between predators and prey animals. Specifically, studies that involve manipulation of animal populations via controlled harvest—made possible through sport hunting—further the understanding of population dynamics because they allow researchers to evaluate whether it is the predators or a different cause that limits population growth of game species. Studies like these primarily reflect the common assumption that harvest by humans and predation are analogous in many ways for their impacts on prey animals. Therefore, a significant amount of funding and other resources support modeling game populations and predator-prey relationships, as these are the prey animals, including deer, moose, antelopes, who, hunters are most interested in. These are also the type of studies that trophy hunting organizations such as the Boone and Crockett Club and the Safari Club International are interested in and provide funding for, which then can lead to the increased killing of predators and even cause them the loss of ESA protection, for example the Boone and Crockett Club has celebrated one of their studies being successful in removing the Eastern timber wolves from Endangered Species protection, allowing that species to be hunted.

Critique: Adaptive (Harvest) Management

Adaptive management is a decision-making process that identifies a specific goal but allows for different paths to get there based on new knowledge, which in turn can inform policy development. This approach does acknowledge uncertainties of knowledge about complex systems as well as uncertainties related to the potential outcomes as a result of different strategies. It attempts to find the right balance between the best short-term outcomes based on current knowledge and gaining knowledge to improve management in the future. Due to its inherent flexibility and involvement of human stakeholders (those who may be affected by project outcomes), it has been successfully applied in forestry management as well as in international affairs where it is used to implement development programs. It is also widely used for knowledge-gathering to inform wildlife management policies. Here, it can be valuable for testing different strategies to improve the lives of wild animals by increasing habitat, food and water access for wild animals, or by trying to find the most effective way to construct channels in order to increase the survival of key ecosystem species, such as the salmon fly.

However, adaptive management strategies have been broadened and used in many instances as adaptive harvest management, in order to test different strategies to meet the goal of maximizing hunting opportunities. As mentioned in Chap. 3, the amendments of the Migratory Bird Treaty Act of 1918 recognize sport hunting as a legitimate use of a renewable migratory bird resource. In 1995, the US Fish and Wildlife Service

(USFWS) adopted adaptive harvest management to provide a framework for regulating harvests of ducks and other waterfowl, as well as for migratory birds. Migratory birds use four major migratory routes (Pacific, Central, Mississippi, and Atlantic flyways) in North America as they migrate between their nesting sites and wintering areas. 'Game' birds in these flyways are heavily managed by the USFWS and partner agencies in Canada and Mexico primarily for hunters, who shot more than 7.3 million doves during the 2016–17 hunting season alone.[17]

Officials of state wildlife agencies in Texas and Missouri, among others in certain regions of the United States, go to great lengths to manage small areas of sunflower fields near urban areas that attract birds and are also convenient for shooters to access. At some point, a Missouri agency-related report explains, 'uncertainties' were identified in regard to the length of daily shootings (half-day versus all day), the quantities of doves to be shot (large daily numbers in relatively short periods versus small numbers spread over a longer period of time). A subsequently established program Adaptive Resource Management (ARM) has identified its overall goal as learning how different management strategies impact the objective of maximizing dove hunting opportunities in public areas. Part of the monitoring programs include the capture and banding doves by the thousands, hunters reporting the number of doves killed, shots fired, hours hunted, zip code, and the number of doves shot but not retrieved. In addition, wings from shot birds are collected to determine the age ratio. This method does not seem to be used to assess strategies to benefit doves. Bag limits and length of shooting times per day are assessed in order to meet the goal of a maximum number of birds that can be shot. This is done to improve the hunters' experience and satisfaction.

NAM's Claim That Predation by Sportsmen Is Equal to Predation by Natural Predators (as Part of 'Harvest Theory')

Inherent in manipulating wild animal populations and their habitats, is an assumed analogy between sportsmen and predators. This is more an expression of the Cartesian view than reflecting the complex reality of natural ecosystems and human society. One of the profound differences is that with limited or no human interference, natural predators achieve a balance (equilibrium) with their prey animal populations so that both persist, while humans have the potential to disrupt these ecological relationships. For example, the rate and intense exploitation of animals by human hunters is not comparable to natural predators. Furthermore, humans also do not mimic the behavior of predators that prey animals are familiar with. In their article, "The Unique Ecology of Human Predators," the authors' findings revealed that humans kill adult prey animals at median rates that are 14 times higher than natural predators do, with especially intense persecution of terrestrial carnivores. This caused the authors to call humans an 'unsustainable super predator,' that will continue to change ecological and evolutionary processes globally unless constrained by wildlife managers.[18]

Furthermore, in some cases, the equilibrium between natural predators and prey animals may move in cycles with some exploding and collapsing. This is the case with the lynx-snowshoe hare cycle in boreal forests. Studies show that just *like* human hunters, lynx move on to less-favored targets when their prey populations collapse, however, lynx cannot sustain themselves on secondary prey for long, and their population subsequently collapses, with emigration and losses impacting their capacity for reproduction.[19] This is not the case with sportsmen, whose population will not collapse. Many of them will simply switch to target different animal species.

Most carnivores hunt to survive and adjust their behavior depending on their needs and on the availability of prey animals. In stark contrast, sportsmen commonly kill animals with significant features such as large body size, big antlers or horns, and other features conveying strength, and these are not the animals that predators usually kill—the weak, sick, slow animals. The same idea applies to trapping, where generally healthy animals are healthy, traveling, hungry and looking for food and become victims of baited traps and snares.

Most importantly, because predators such as wolves, mountain lions, and bears are so much more than just 'killers of prey animals,' sportsmen cannot substitute for the ecosystem services of the animals they kill. Sportsmen may extinguish an individual animal who is best adapted to her/his environment. Finally, sportsmen cannot replace an animal lost to her/his mate or family members that will be missed and grieved over by surviving animals. How could sportsmen possibly replace the mother bear who is still nursing and taking care of her cubs? She is their only lifeline and without her, the little ones will likely die as well.

The Role of Science and Ethics in Wildlife Policymaking

In Western society, we use scientific knowledge for the decision-making processes in the industry, institutions with wildlife policymaking being no exception. Science can help inform us about the choices and consequences of our actions. It can even give us insight into what actions produce better or worse outcomes in the world, and these insights may help form better policy.

Scientific reasoning and evidence help to distinguish better from worse causal explanations, and they can expose myths. Animal ethicist scholar, William Lynn notes, "… scientific knowledge is always contingent, and laden with values, worldviews, and vested interests."[20] Sportsmen have vested interests in the continuation of hunting and trapping, both of which require the survival of wild animals and their habitat. This requires management and the creation of an additional number of 'expendable' wild animals—the creation of a "shootable surplus" of animals. A huge infrastructure system around state wildlife agencies, and resource allocation, including research, within and by these agencies serve this goal.

Adding to the high stakes involved in animal exploitation and associated truth claims (see Chap. 2), professor of environmental studies at the University of Montana and my former doctoral advisor, Dan Spencer has cautioned that when evaluating scientific claims, close attention should be paid to three components: first, it must take a critical rather than reductionist view of describing claims for truth, guided by the awareness that scientists' own values and assumptions are inherent in choosing what is important and how to interpret findings. In other words, as much attention must be paid to what is hidden and not revealed as to what is claimed to be discovered and uncovered. Second, one needs to keep in mind that the important role of science should not be drowned by pluralistic relativism. Instead, it should be acknowledged that science plays a central role in understanding the world and the most important question that emerges here is asking how science can play a constructive role as part of a liberationist ethic. Third, scientific knowledge must be linked with other ways of knowing that inform and guide our praxis. "Separating scientific knowledge from other ways of knowing results in a reductionist epistemology characteristic of the Enlightenment mechanistic worldview—an epistemology that threatens the diversity of experiences and meanings in our social and ecological worlds."[21]

As any practical ethics for conservation must recognize the moral value of animals,[22] Chap. 2 has argued that the experiences and ways of knowing by other animals—the Animal Standpoint perspective—need to be included as part of the best available science to help develop policies guiding our human relations with the more-than-human world. This would mean that, in recognition of justifications for protecting human subjects reflected by human rights and civil liberties, so too should conservation recognize and act on the possession of rights and civil liberties of every individual wild animal and be motivated to develop positive relationships with other beings that protect and strive for their well-being and flourishing.

As a method of discovering and testing knowledge—albeit just one, science is one of the most powerful forces in modernity in providing information that shapes philosophies and ethics, and scientific practices and assumptions have significantly influenced much of human thinking and relations, including our view of other animals and the rest of nature.[23] While science can help inform us about the choices and consequences of our actions, which in turn might lead to better policy, we need ethics in order to make moral choices about how we ought to live and have ethically positive relationships with others. Science and ethics are indispensable to, and complement, one another. However, as Spencer has further cautioned, "[S]cience as a discipline aims for objective accounts to describe and explain the world, yet the accounts it gives must never be confused with the reality they seek to describe. Science always remains a human discipline and discourse that takes place within and draws upon the conceptual and value frameworks of its historical and social context." In addition to perpetuating cultural and intellectual assumptions, science also incorporates and reproduces societal dynamics of power.[24] In the case of the North American NAM for Wildlife Conservation, this is mostly to the detriment of wild animals, who suffer injuries and lose their lives for human use and recreation.

It is common among wildlife professionals to believe that their use of science is value-free and that belief systems and values held do not influence management decisions. It is equally common for wildlife or furbearer managers to refer to the principles of conservation, while simultaneously claiming that these values are not important in management decisions. Yet, every management decision and every principle formulated in wildlife and furbearer management is based on a value judgment, which has ethical implications. This type of false separation of science and value takes us back to the 2013 analysis of the NAM by Vucetich and Peterson (see above), who pointed out that the claim that hunting is supported by the best available science, is a misrepresentation of science, as there is no science that concludes it is necessary to hunt [and trap] wild animals.

When lethal use of other animals is the goal of scientific wildlife management,[25] wild animals are treated instrumentally and literally as nonfeeling things, bearers of fur, and props for the human experience of having hunted or trapped, to be used by humans. Thus, according to the NAM's worldview and most prominently expressed in the literature on trapping, management, and conservation of furbearers, these wild animals are merely valued by NAM's science for their production of fur akin to the agricultural NAM that views and treats sentient cows, pigs, and other livestock animals as mere food production units. This view still echoes Aldo Leopold's words in *Game Management* in which he compared game management to "other agricultural arts" that produce a crop..." (1933, p. 3).

Because of the NAM's overall view of wild animals as nonsentient things without emotions, cognition, or subjective experience, it is reasonable to assume that this shared belief and the shared history and management goals, by hunters and trappers led to ready incorporation of recreational and commercial trapping of wild animals within the NAM's own developing structures and principles. The NAM's objectifying view of wild animals as natural renewable resources in general and game animals as food and protein units in particular, conceptually meets another commodifying practice that views furbearers as "fur producing units" on the same level, notwithstanding the conflicting consequences this has been posing for the NAM, such as the issue of commercialization that trapping represents. Similarly to other large-scale industries exploiting animals, for example, for food, clothing, and biomedical experimentation, so too has the industry of wildlife profession been successful in turning animals and their lives into numbers, statistics and ecological units, producing value for the human experience, to be hooked, shot, and trapped. The NAM's selective use of science has effectively and efficiently erased animal individuality and subjectivity in order to facilitate human interests. As a pro-trapping brochure by the Northeastern Committee of The Wildlife Society explains, "individual animals do not matter, only species and populations count."[26] However, this statement has little to do with science but all to do with value. According to this type of management by numbers, wild animals and their lives are merely represented by numbers, statistics, and quota, and only their lives, or rather, their deaths, have a place only in wildlife management's accounting books. However, the late wolf biologist Gordon Haber objected a long time ago to this type of management by the numbers and emphasized that a healthy wolf population is more than x number of wolves inhabiting y square miles of territory. The notion that we can 'harvest' a fixed percentage of a wolf population corresponding to natural mortality rates and still maintain a viable population misses the point. According to Haber, it's not *how many* wolves you kill, it's *which* wolves you kill.[27] In the same vein, scientists studying other species, have discovered that individuals often play various and significant roles within their own communities. For example, by applying research techniques used to analyze human social networks to study the social networks of free-ranging bottlenose dolphins living in Doubtful Sound, New Zealand, Lusseau and Newman (2004) found the following: a) communities and sub-communities exist within dolphin populations, b) sex-and age-related homophily played a role in the formation of clusters of preferred companionship, and c) so-called centralized brokers, certain individual dolphins exist, who are at the boundaries between communities, appear to be crucial in maintaining the cohesiveness of the dolphin community. The latter may be regarded as "key individuals who can control the flow of information in the society"[28] (p. 479). The importance of individual animals playing essential roles and the need for their presence within their own communities has been further supported by research that found that lethal control of gray wolves backfired and actually increased depredation; findings showed that killing one wolf increases the odds of depredations 4% for sheep and 5 to 6% for cattle. If 20 wolves are killed, livestock deaths double (Wielgus and Peebles 2014). Their study did not confirm the "widely accepted but untested hypothesis" that killing wolves leads to less depredation. Instead, the researchers concluded killing wolves disrupts the social cohesion of wolves' society and sets in motion a downfall where "disruption can set sexually mature wolves free to breed, leading to an increase in breeding pairs. As they have pups, they become bound to one place and can't hunt deer and elk as freely. Occasionally, they turn to livestock" (Sorensen 2014).

The NAM's human supremacy worldview is manifested in its selective use of science that also locates humans as outside and above nature, assuming the role as its manager (the subject of knowledge holder) and treating other animals as objects (non-feeling things) and commodities can then be used as means to a human end. As such, traditional science used in order to facilitate exploitation of wild animals also entails powerful decision-making of what and who counts, who, when and how nonhuman animals die. To this end, hunting and trapping seasons are set to regulate the annual killing spree for sportsmen, but the killing methods used to end the lives of trapped animals are not. Traditional science here makes a clear statement that this is unimportant.

How Proponents Justify Trapping

NAM proponents consider "regulated hunting and trapping as the cornerstones of wildlife management."[29] Arguments put forward by trapping proponents consistently describe trapping this way: "Government-regulated trapping in North America is ... a versatile, safe, effective, and ecologically sound method of harvesting and managing furbearers ...

trapping is part of our cultural heritage that provides income, recreation, and an outdoor lifestyle for many citizens through the use of a renewable, natural resource. The section below discusses trapping as heritage/tradition and sport/lifestyle.

Trapping as a Cultural Heritage/Tradition

The selectivity of the Model's narrative of the man-the-hunter myth has been discussed above, and the same line of argument applies to trapping. Also, at the time when the state institutionalized trapping, it acted so on the moral insights of its citizens. The position statement by The Wildlife Society claims that we should continue the practice just because of the length of time trappers have indulged in that activity. However, the content of a tradition, not the length of duration, should be the deciding factor in whether to continue or discontinue a tradition. Nelson et al. (2011) have explained that, "[T]o suggest that a historical episode can justify an ethical prescription is to commit a logical fallacy known as *argumentum ad antiquitatem* (the argument from antiquity or from tradition)" (p. 59). For example, Western society has deemed certain past traditions, including slavery, wife beating, and female circumcision as unethical and decided to end these morally unjustifiable practices. Likewise, to argue that trapping should continue just because it has played a role in the past is logically and ethically wrong.

Tom Regan declares that appeals to retain this tradition of trapping "themselves are symptomatic of an impoverished view of the value that animals have in their own right and thus can play no legitimate role in defending a practice that harms them."[30] In her exploration of subsistence traditions by indigenous peoples in North America (First Nations), involving killing of nonhuman animals, Montana-based theology scholar and animal rights advocate Lisa Kemmerer (2004) elucidates the meaning of tradition, and compares contemporary lifestyles and practices with traditional ways. Most contemporary activities, Kemmerer argues, do not qualify as traditions anymore as they are either missing one or the other components that make up a traditional activity in the true sense. Consequently, these activities should be described as *new* traditions, which may no longer qualify as worthy of protection to continue. According to Kemmerer, two aspects are critical of a true tradition—a continuing pattern and methods, including a belief structure. Kemmerer elaborates further by stating, "methods entail an inner quality; a mind-set—the way people *feel* about and *envision* a particular activity."[31] The primary purpose of indigenous peoples engaged in hunting, fishing, and trapping was survival. The conflict between wildlife as kin and wildlife as being critical to survival presented a constant conflict between "the necessity to eat and the killing of animals, which [were] believed to possess consciousness, will, and soul" (p. 3). This tension resulted in a spiritual relationship guided by religious beliefs and ritual acts, which alleviated the guilt and kept the killing in check. At the same time, Kemmerer explains that this core wildlife ethic of some of the indigenous peoples, which included fasting, prayer, and avoiding taboos, also viewed animals not only as powerful physical and spiritual beings but as teachers and guides who could communicate with people.

Contrary to this spiritual subsistence killing, most early white European trappers were motivated by commercial incentives right from the beginning without any spiritual relationship with animals. Without a spiritual connection to the animal world that may have kept the killing force in check, trapping for commercial profit led to an uncontrolled slaughtering of animals, of which some species were brought to near extinction, such as beavers. Though the main reasons for historical and modern trapping, (financial profit and an outdoor, frontier experience) have remained vaguely similar, the methods and the mindset have changed and trapping has become a new practice. Where early white trappers traveled by boat, horse, or foot, the modern trapper gets to the trailhead by means of a 4×4 truck and then switches to a snowmobile to check his 200-mile long trapline. This modernized process of trapping stretches the term tradition to the breaking point because of radically different methods (physical means) and demonstrates that trapping has transitioned into a recreational activity along with a new mentality. Trapping cannot be called a tradition just because one's ancestors trapped, or because one sets traplines in the same area and at the same time of the year as one's ancestors. Modern trapping, stripped of its historic purpose as a way to survive[32] and certainly absent of a spiritual relationship with animals, does not qualify as a historic tradition; rather, if it has to be called a tradition, trapping should be categorized as a *new tradition* at best. According to Kemmerer's definition, a new tradition is "an activity linked to historic traditions but largely composed of methods developed in the past century and/or lacking core spiritual or philosophical aspects that once accompanied those actions."[33] A new tradition needs to undergo scrutiny for its appropriateness in modern times: Society has implemented animal cruelty laws to prosecute people who knowingly cause pain and suffering to animals. Since this is exactly what trapping entails, this new practice cannot be justified ethically.

Trapping as a Lifestyle

Another aspect that invokes the heritage argument in attempting to legitimize trapping is that it often refers to a lifestyle choice of trappers. This generally refers to a sociocultural study by Daigle et al. (1998) of trappers, 98% of whom were male and 98% were white, in six northeastern states, which found that, in contrast to previous studies, trapping was a "well-integrated" activity "within a broader lifestyle oriented toward the out-of-doors." The study, which states at the beginning of its introduction that "trapping provides intense psychological and emotional satisfaction," showed that not purely economic reasons, but broader social and cultural motivations underlie participation in trapping wild animals, first and foremost as a component of an outdoor lifestyle that includes "hunting deer, upland birds, and small game; fishing; gathering wild, edible plants; cutting firewood for household use; and planting vegetables" (p. 622). This primary lifestyle factor was followed by "nature appreciation," or a view of helping/managing "animal control" (p. 622). The problem with arguing for trapping in terms of heritage or lifestyle protection is that it involves killing of animals not out of need such as self-defense or for subsistence (as a last resort in an extreme situation), but as a lifestyle *choice*. This human choice "involves trade-offs of vital animal needs for trivial human interests"[34] as the trapper's lifestyle choice comes at the expense of another animal, an animal who undeniably has an interest in staying alive, and an animal who does not pose a threat to anyone's life. Thus, in this context, "[K]illing the animal becomes, in effect, entertainment, a means of escaping the pressure and rigors of civilized life" (Mallory 2001, p. 65).

The appeal to tradition will work only if it promotes continuity of ideals while presupposing common needs and purposes. This, however, is no longer the case—the frontier is gone and reasons relevant to contemporary circumstances are required to justify trapping, an activity of some that is simply unacceptable for anyone who also feels connected with other individual animals. In summary, as pointed out by Canada-based scholar Margaret Van de Pitte (2003), a traditional practice may be lacking morality, thus an independent argument for moral legitimacy is required to establish it as moral; secondly, and applying Van de Pitte's reflection on the right to (sport) hunt to trapping for recreation, while one can plausibly go from affirming the right to subsistence to the right to hunt (disregarding context), one needs additional grounds to go from the legitimacy of subsistence trapping to that of trapping for fun. In summary, the democracy of trapping no longer works in a society that increasingly finds trapping for recreation and commerce morally repugnant. And, however gratifying trappers may view their practice of trapping animals for their fur, their pastime nonetheless involves a willingness to override the animal's vital need to survive and be unharmed by humans, in order to promote the experience of having trapped. The moral issue is glossed over by The Wildlife Society and other promoters of the Model, but the question remains: What right do humans have to deprive an animal of her/his life if we do not need to do so?

Trapping as Sport/Recreation

The claim that trapping is a sport lacks any substance. Trapping does not meet the definition of sport as "an activity that is governed by a set of rules or customs and often engaged in competitively. Sports commonly refer to activities where the physical capabilities of the competitor are the sole or primary determiner of the outcome (winning or losing)," (Wikipedia). Certain ethical criteria are requirements and prerequisites for our understanding of a fair sport, such as tennis, which, when played by competing individuals include a) both participating parties engaging in the activity voluntarily, b) both parties knowing and playing according to the rules, and c) both players having similar physical abilities. Trapping is just the opposite of all criteria defining a fair sport. The notion of fairness as part of the sportsman's ethics code in hunting, does not exist in trapping when trappers deceive sentient and unsuspecting animals——targeted and non-targeted animals——by luring them into concealed and baited devises that often injure the caught animal. In contrast, when wild animals engage in play behavior with their family or pack members they know the rules. But, animals slated to be trapped do not know any of the trappers' regulations or rules nor do they know that they are being used as playmates in a lethal game, which has a deadly outcome only for them and not the trapper (significantly unfair). Trapping wild animals for their fur is a great injustice and reveals callous behavior with an instrumental attitude toward animals, where trappers gain positive values in the form of fun, excitement, outdoor experience, and financial gain—— while the captured animal is left for an unknown period of time to struggle in the trap where it experiences pain and suffering and subsequently pays with its life. As for the outdoor skills trappers claim to be able to develop only through trapping wild animals, taking photographs of these animals would serve the exact same purpose, particularly if one wishes to photograph rare species such as wolverines, fishers, and swift foxes. The question becomes, what establishes the practice of trapping living and feeling wild animals for their body parts as a heritage to be continued rather than merely remembered?

Also, contrary to hunters who have developed an ethical code for killing animals that declares that the goal is fair chase and a clean lethal shot to minimize the pain and suffering of those animals they pursue for sport, trappers have never moved beyond a respect-and-do-not-waste-the resource attitude. By respect, trappers mean respect for the pelt, not the animal. Ecofeminist and animal rights advocate Brian Luke (1997) has argued that "the 'fair chase'—refraining from killing animals who cannot escape due to water, deep snow, fences, chasing with a vehicle, electronic tracking, automatic weapons — is superficial fairness as it is not in the interest of the animals to be hunted and killed in the first place" and, "not a balanced, jolly game of mutual satisfaction. I get to shoot you and you get to be dead" (p. 27). However, there is not even the slightest trace of even superficial fair chase inherent in trapping: trappers place concealed traps and snares on well-known paths that the animals use to travel across the landscape. Furthermore, they use baits—often the carcasses of small animals they have trapped and skinned—to lure animals into lethal traps, where they are left to struggle, suffer, and die.

Trapping deceives animals because traps take animals by surprise without granting them a fighting or ethical chance; besides being injurious and deadly, traps are a novelty to wild animals because a) as opposed to other predators or other natural occurrences, traps are outside of things and situations they can anticipate or expect in their interactions with the environment, and b) most animals cannot adapt to traps and learn to avoid them in the future since there is no future for most animals caught in a trap. Instead, getting caught in a trap means the occasional quick or more common, agonizingly slow death, caused either by the trap itself, by force from the returning trapper, or, in cases where an animal escapes by chewing off its foot, by later predation since death for a three-legged animal is often imminent. The trickery applied in trapping goes even further. A former trapper told me that his trapper friend had just called him to brag about his victory. He'd been trying a female coyote for three years. She had always outwitted him, knowing the dead animal parts he put by his traps were suspicious lures. The trapper hit on an idea and found her den. He grabbed and killed one of her pups and smeared the scent on his trap. Searching for her missing pup, the coyote mother found the scent, stepped in the trap, and the trapper killed her. He was so proud to finally have proved he was cleverer than she, bashing her head in until she died, he had to call his friend and brag about it. It is more than likely her other pups died of starvation.[35] The trickery inherent in trapping has been criticized by philosopher Paul Taylor who opposes the deception with the intent to harm and notes that the betrayal comes in form of "knowingly taking advantage of an unsuspecting animal and thus treating the animal as a mere means to an end..." Taylor argues further, "being deceived in order to bring advantage to the deceiver reveals the deceiver's view that the animal has either no inherent worth at all or a value lower than the deceiver himself."[36]

Textbooks in wildlife biology, such as Bolen and Robinson (2003) acknowledge the self-regulation capacity of populations of animals. Other publications note that healthy populations such as muskrats cycle throughout the year. The annual increase and decrease is the natural, inherent health modus, which furbearer management takes away and replaces with artificial management of these animals. All arguments flow here, that is, that trapping is a needed tool to reduce the number of animals within a population because many diseases (e.g., sarcoptic mange and rabies) are density-dependent, is unconvincing for three reasons. First, trapping may, in some cases, cause a spike of animals above their habitat's carrying capacity thus, increases density; second, disease outbreak, whether or not density-dependent, is a species-specific mechanism of density reduction; third, arguing that trapping is more humane than starvation is a mere value statement made by wildlife agencies and trappers, and certainly cannot be used to argue that trapping is required.

Trappers are mostly interested in the prime fur of healthy animals that fetch the most dollars at fur auctions. It is obvious that trappers disregard the welfare of the animals they capture and kill because they use leghold traps that have been declared inhumane by many animal-health associations and they continuously oppose any regulatory changes toward a mandatory trap check. With these facts in mind, it is clear that trappers lack empathy for the animals they pursue. Thus, trappers' arguments in favor of one practice over the other are without merit.

The significant physical and psychological trauma that animals experience, particularly in restraining traps will be discussed in Chap. 7. Also, starvation, as unpleasant as it may be, is *nature's tool* to eliminate the weak and sick animals and help ensure that healthy and strong animals survive. A natural balance of predators to prey diminishes the chance of death by starvation. Trappers, on the other hand, do not seek out sick animals to kill; instead, they trap mostly healthy, curious, and moving animals. Furthermore, trappers frequently state that they kill animals in traps more humanely than other predators. This assertion ignores the fact that as moral agents, "we make moral choices according to principles of right and wrong and need not operate simply by instinct in our eating behavior."[37] Rarely, if ever, are animals trapped for food today. The claim that trapping is more humane than predation assumes that traps do not catch predators, when, in fact, most furbearers are both predators and the target species. This argument is simply a manipulation, a misleading statement meant to divert attention from the indefensibly cruel practice of trapping.

Finally, according to the trappers' claim, they perceive themselves as instruments of nature and their violent acts against animals merely reflect the violence of the natural world. Trappers view themselves as better than nature by killing animals in a more humane way than nature. As such, trappers claim to provide a humane service to animals, an activity they perceive is beyond ethical reproach that should be embraced by the public. How humane is death for an animal in a trap? A trapped animal unfortunate enough to be alive after spending days and nights struggling for freedom, successfully or unsuccessfully fending off predators, suffering without water and food, and surviving in freezing temperatures, will be either clubbed to death, suffocated, or shot. The terror and pain an animal in a trap must endure before being put out of misery can hardly be called humane or kinder than a natural death. The trappers' argument does not hold water, especially in a state like Montana where no trap check is required. Here, trapped animals can linger in traps for an unknown period of time and endure just what trappers claim to prevent: starvation and predation in addition to the pain and suffering caused by the trap.

The NAM: Negative Emotions and Old Beliefs Drive Policies

Sportsmen often provide emotional testimony about how important hunting is as a personal experience, engagement, and even for their mental health. The emotions and values lead to policies that in turn inform the grounding of laws. Informed by science, a policy constitutes a course of action chosen from among several possible alternatives as a guide to determining present and future decisions. A particular policy is not only a product of the political process but reflects the distinct culture in which it is formulated and operates. Today's policies guiding wildlife management and conservation programs by governmental agencies reflect the traditional culture of sportsmen—a narrow value system of a minority of approximately 6 percent of the American public—whose interests in killing animals have not been scrutinized by the general public. State wildlife policies are an expression of the contemporary (and historic) culture of how wildlife professionals—and the sportsmen who lobby them—think, their view of wild animals and actions toward them. As one of the NAM's own publications explains, contrary to scientific findings, sportsmen continue to believe that eliminating upper trophic level carnivores such as wolves, bears, mountain lions, and others, will inevitably result in more deer and elk to hunt. In the same vein, the belief is prevalent among upland game bird hunters, and thus, raptors and medium-sized mammals including raccoons, opossums, foxes get shot in order to greater hunting opportunities of quail and other birds.[38] In light of these misguided but prevailing assumptions, the NAM has clearly not succeeded in educating its own constituency about the value of science. The fact that the human supremacy worldview along with traditional and conservative values are shared by most agency personnel may explain this failure, yet the North American NAM's tenets are committed to science-based policies. That wildlife management has been dominated by sportsmen's values that have not much to do with the best available science is currently playing out in states with wolves present with tragic consequences for these animals.

The NAM and the Relentless Persecution of Gray Wolves: Politics

Gray wolves (*Canis lupus*) have been persecuted in the U.S. since the arrival of Europeans. In the mid-1990s, wolves were reintroduced in Yellowstone National Park (YNP). By the 20th century, they had been driven to near extinction. In the 1970s, wolves in the United States were rescued from the brink of extinction in the 1970s and protected under the federal Endangered Species Act (ESA).

In April 2011, gray wolves in Idaho, Montana, eastern Washington, eastern Oregon, and northern Utah lost their protection under the

Endangered Species Act. Their delisting was not the result of scientific consensus but was forced through pressure by sportsmen and the livestock and ranching industry. It was accomplished through a political maneuver by U.S. Sen. Jon Tester (D-MT) and U.S. Rep. Mike Simpson (R-ID), who attached a last-minute rider to a must-pass budget bill, that stipulated the delisting of wolves. This move also blocked any judicial review of the rider, which in turn prevented any legal challenge to the wolf delisting (so much for a democratic process).

Previous decisions to delist wolves in the Northern Rockies in 2008 and 2009 were legally challenged by conservation organizations, which halted the start of hunting of wolves. It seems that states have responded to each relisting effort by the public, with even more liberalized hunting and trapping policies to eventually persecute wolves. While the public is increasingly revering wolves and petitioning for their protection, under the North American NAM, they and other predators continue to be perceived as a threat to hunting opportunities, economies, and despised by ranchers and the corporate livestock industry.

The Case Against State-Management of Wolves: Example Idaho

On May 6, 2021, Idaho Governor Brad Little signed a bill into law that allows sportsmen, and even paid private contractors to shoot and trap up to 90 percent of the wolves living in the state. The new law permits the killing of wolves by various cruel and highly unethical means including night hunting with night-vision equipment, aerial gunning, chasing and shooting wolves from snowmobiles and all-terrain vehicles. In addition to allowing individual hunters and trappers to kill as many wolves as they want, the new law authorizes year-round wolf trapping on private lands, including during the season when pups and females are most vulnerable. The state also allows a $1000 "bounty" paid to trappers per dead wolf, including wolves killed deep within some of America's largest and wildest Wilderness areas.

In the past, Idaho's state management of wolves has included Idaho Fish and Game (IDFG) hiring a professional hunter-trapper to go into wilderness areas to slaughter wolves and IDFG conducting aerial gunning operations to slaughter wolves in some of the most remote roadless federal wildlands remaining in the lower-48 states. Records obtained from IDFG show that at least 570 wolves were killed in Idaho between 2019 and 2020 with abhorrent brutality.

At least 35 wolf pups, some weighing less than 16 pounds and likely only four to six weeks old were among the killed[39]… one wolf caught in a trap had only three legs with a trap still stuck on one leg… two wolves had cracked teeth to the bone or shattered teeth trying bite traps…. one wolf was found dead in a private trap and one wolf trapped by USDA Wildlife Services in July died of hypothermia, while several other wolves were gunned down by Wildlife Services in "Aerial Control Actions."[40]

Other state-ordered extermination actions include IDFG paying $22,500 for the aerial killing of 14 wolves in northern Idaho, hiring professional hunters/trappers to kill wolves, even entire wolf families living deep in the Frank Church-River of No Return Wilderness, some with the declared purpose of increasing elk populations for outfitters and recreational hunters. Despite public outrage and several lawsuits filed to try to intervene and stop the slaughter of wolves, at least nine wolves were killed by the professional killer as of January 27, 2014.[41] That same year, and with the permit from the federal US Forest Service, IDFG landed helicopters in the wilderness in order to track and collar wolves. This was the first step in a plan to eliminate as many wolves as possible in the wilderness. In addition, between July 2017 and May 2018, the federal USDA WS killed at least 53 wolves in ID to "avenge livestock depredations" at the tune of half a million dollars of taxpayer money.[42] Trophy hunting organizations provide additional money, for example, the Rocky Mountain Elk Foundation (RMEF), which provided $350,000 in grants specifically for wolf management" over the course of three years, prior to 2015. Since 2016, the RMEF has provided significant financial support to the misleadingly named Foundation for Wildlife Management in Idaho to help fund a 'bounty program' that pays between $500 and $1000 for each wolf killed by a hunter and a trapper respectively. The RMEF emphasizes that wolf/predator management is part of the NAM.[43]

The Case Against State-Management of Wolves: Example Montana

In February 2021, Montana's governor Greg Gianforte learned a black wolf was caught in one of the leghold traps that he and another sportsman had set. Being near Yellowstone National Park where his wife wanted to shoot a bison, Gianforte had to travel to the trap site and upon arriving, he got to shoot the trapped wolf—an illegal act as it turned out. Outfitted with a research collar, "Max" as he was known, had just come from Yellowstone National Park and crossed the park border, probably in search of a mate and/or territory.

While Gianforte called shooting the trapped wolf "an honor," the editor of a Montana newspaper wrote that to him, "the integrity of the moment was hardly different than shooting a chained dog."[44] Gianforte did not care that Max, who had just stepped across the protective order of Yellowstone, was wearing a research collar that held valuable scientific research data about the movements of park's wolves. This is not atypical for sportsmen, for example, in 2012, FWP had to temporarily close the hunting season on wolves in units around Yellowstone because sportsmen in Montana, Idaho, and Wyoming had already shot seven collared wolves coming out of the park—Yellowstone is home to approximately 85–100 wolves. At the end of 2011, hunters had shot a combined 191 wolves out of a region-wide population of at least 1774 animals.[45]

A dedicated hunter and lifelong trapper, Gianforte also did not have yet a certificate of the wolf trapper education class, a legal requirement before one can legally trap wolves in Montana. However, the Governor merely received a slap on the wrist by the department of Fish, Wildlife and Parks, an agency he happens to oversee as Governor.

Then in spring 2021, Montana's approach changed when its legislature passed an avalanche of anti-wildlife and anti-wolf bills, and essentially issued the state Fish and Wildlife Commission two mandates: increase "harvest" and legalize suffocating neck snares for killing wolves. Since then, wildlife management has changed dramatically and taken a turn for the worse thanks to the sportsmen-dominated legislature, a trophy-hunter-influenced Commission, and Governor Gianforte, a lifelong sportsman. That Commission, all but one appointed by Gianforte now represents such anti-wolf industries as the Montana Stockgrowers Association, the oil and gas industries, and the Montana Outfitters and Guides Association, and one member from the world's wealthiest and most notorious Safari Club International, a trophy hunting organization (see Chap. 3).

Signed by Governor Gianforte into law, the new law now extends the wolf killing season to six months each year, quadrupled total bag limits from 5 to 20 wolves per person (ten wolves can be killed through shooting and ten wolves can be killed through trapping), and removed buffer zones around Yellowstone and Glacier National Parks. They allow a "bounty" paid by an anti-wolf nonprofit organization to each sportsman who kills a wolf. legalized strangulating neck snares, which, strategically placed, can capture and strangle an entire unfortunate wolf family (a so-called "pack") at once, as well as night hunting, spotlighting, and baiting—"poachers' tools" as the only dissenting Commission member, a hold-over from the past era of wildlife/wolf management called them.

Intensifying the Slaughter of Wolves Against Science and the Public's Will

This new era of anti-wolf policies has been met with legal challenges, and the public expressed their rejection in more than 90 percent of the 26,000 comments sent to Montana Fish Wildlife and Parks against these extreme killing policies. That disbelief and rejection also came from seven former Fish Wildlife and Parks Commissioners who called the new regulations "regressive," "egregious," "zealous," "irresponsible," "unethical"—an unprecedented condemnation of a Commission's successors.[46]

In addition, a letter signed by a group of Montana wildlife biologists and managers was published who are worth quoting: "Wolf haters in the legislature [have] made up fact-free stories about the impacts of wolves on elk populations in order to destroy Montana's recovered wolf population by legislating…the use of methods [that] are completely contrary to the accepted principles of fair chase hunting and hunter ethics."[47] What provoked them? A sense of moral outrage. And the legitimate fear that, with this coordinated effort to destroy in whole or at least in large part an entire region's wolf population, wolf numbers will quickly drop to zero.

Selling wolf kill licenses is a very lucrative business for state wildlife agencies. For example, Montana's FWP sold 14,921 resident wolf licenses and 2082 non-resident wolf hunting licenses during the 2018–19 season, which generated $387,599 for the agency.[48] However, compare this to Yellowstone wolves, who generate between $60 and $70 million tourist dollars annually.[49] But not even the powerful economic force by wolves saved them from slaughter as FWP and other state agencies do not discriminate between a wolf who crosses the park's protective border and steps into a minefield of sportsmen lying in wait to shoot and trap them or wolves in any other parts of the state.

Wolves from Yellowstone National Park

People come from all over the world to catch a glimpse of wolves. On any given day throughout the year, groups of tourists spend hours on end behind telescopes to watch wolves' behaviors and antics. This also means that Yellowstone wolves are rather used to the presence of humans and it can be assumed that wolves do not suspect any danger coming from a sportsman lying in wait outside the park's border. Sadly, even when hunters lift their arms to point their rifle toward wolves, these animals will likely not be suspicious of that movement before they get shot. What does this have to do with 'fair chase'?

Yellowstone National Park shares borders with Montana, Idaho, and Wyoming, the three states that have the most heinous wolf killing policies. These three states have the most heinous policies and seemed to be determined to maintain only the minimum number of wolves needed to avoid relisting of wolves under the ESA. This bare minimum number is 150 wolves embodying 15 different packs in each of the three states and was set by USFWS as a requirement for delisting.

The bleak situation facing wolves in Montana, Idaho, and Wyoming proves that gray wolves still need protection under the federal ESA. The hateful mindset that declared wolves "vermin" in the past and nearly caused their extinction has not changed. Long outdated beliefs and the human supremacy worldview endure throughout time and have not changed, they have certainly not changed among many key decision-makers deciding over the fate of wolves and other wild animals: the hunters, the trappers, the state wildlife agency personnel, the ranchers and associated livestock industry among other industries that rely on the exploitation of wild animals. And no animal species should lose their ESA protection, instead, animal species should only be delisted when the culture of human society into which they are about to be 'released,' has changed toward co-existence and tolerance, none of which exists under the North American NAM.

As the domain of sportsmen, the treatment of wolves in western states, particularly in Montana and Idaho shows a high level of intolerance of native gray wolves, even in light of extremely "lax" regulations and a high killing rate. Before the end of the hunting season in 2019, a news article on March 15 reported 315 wolves killed by sportsmen already in that season alone.[50] Yet even this carnage was not enough for Montana's 'sportsmen,' who called for an aggressive campaign for opportunities to kill more wolves at all costs—including aerial hunt and, reminiscent of the preferred method during the near extermination 100 years ago: poison. "Montana used to be for the hunters, not for the predators!" lamented a sportsman at a public meeting in Montana.[51] As of mid-February, 2022, 223 Montana wolves have been legally killed, including 85 wolves—more than one-third of the Park's—as soon as they've stepped across the Park's invisible boundary line, and an unknown have been illegally killed in Montana as well as in other Western states.

Wolves as Scapegoats

"Let's be honest. What is happening in Idaho and Montana is not hunting. It is pure, unbridled cruelty." These are the words by Daniel Ashe, the former US Fish and Wildlife Service in a 2021 article in the Washington Post, calling for an emergency re-listing of wolves.[52]

Montana's sportsmen, ranchers, and industries hostile to wolves insist wolves will bankrupt ranchers, elk-hunting outfitters, and sportsmen.

But economics and ecological science simply do not support these most prevalent justifications.

If it were about science, thousands of peer-reviewed studies would be considered, all of which vindicate wolves as critical players in healthy ecosystems. The reintroduction of wolves to Yellowstone National Park has shown the transformative power of wolves' presence leading to the restoration of the land and riverbanks and resulting in a significantly high level of biodiversity that benefits beavers, aspen, and bird species among many others.[53] The importance of individual animals playing essential roles and the need for their presence within their own communities has been further supported by research that found that lethal control of gray wolves can disrupt the social cohesion of wolves' society, actually backfired and actually increased depredation; findings showed that killing one wolf increases the odds of depredations 4% for sheep and 5 to 6% for cattle. If 20 wolves are killed, livestock deaths double.[54]

There have been scientific assessments by wildlife biologists that wolves were not fully recovered to meet the baseline standards in the ESA, concluding that former President Trump's delisting in January of 20 was premature and not warranted. "Human-caused mortality sanctioned by state governments," once again puts wolves on the biological edge, they wrote, citing Idaho, Wyoming, and now Montana as hostile states.[55]

Science also acknowledges that wolves have a minuscule impact on ranchers' economies. Here, it needs to be pointed out that after all, it is the ranchers' choice to leave most of the roughly two and one-half million cows and two-hundred thousand sheep, unsupervised on public lands, left to fend for themselves. These animals are mostly simply left on public lands, unsupervised, and left to fend for themselves before being rounded up and send to slaughter. The kill rate of these poor animals by wolves is minuscule from a strict numbers point of view, yet wolves are being blamed for killing cows and sheep at a rate that is detrimentally impacting ranchers. In 2019, wolves killed 69 cattle and 21 sheep. The state reimburses confirmed wolf kills at 100% of fair market value, and in 2019 Montana reimbursed 32 ranchers $82,400.[56] Compare this to freezing temperatures that often kill thousands if not tens of thousands of cows, as happened in 2018, when at least 11,000 cows died a brutal death during a blizzard's freezing temperatures in Montana. After two brutal winters, Montana ranchers asked for an expansion of livestock losses and to be compensated for the deaths of the cattle they caused.[57]

If it were about science, wolf management would consider the science of animal brains, which concurs that wolves have brain structures and capacities that govern consciousness, emotions, and cognition comparable to our own. They have perspectives on the world, individual personalities, and quirks. Their lives matter to them, as do the lives of individuals with whom they have emotional bonds.

NAM's Take-over by Profit-Seeking Extremists Among Sportsmen

Driven by a growing influence by commercial and private profit interests—RMEF, SFI, the Outfitter Association, and wealthy landowners—presents an undercurrent that has produced a flood of even worse legislative bills, and other changes that altogether going into the direction of intensifying exploitation of wild animals at all costs and also toward increased privatization of the killing of wild animals.

In contrast to the NAM's claim that it can withstand any political changes, the developments in Montana and other western states show that the NAM is exactly that—highly vulnerable to extreme hostile legislators and extreme politicians, conveniently joined by industries and decision-making limited to a small circle of government agencies, legislators, landowners, especially landowners and sportsmen.

Sportsmen have used their influence and lobbying power to direct wildlife management, which in Montana in 2021 amounts to a literal take-over of state-administered management, now by commercial and privatization interests. Especially in recent years, the partnership between state agencies and sportsmen has taken an even more worrisome turn, where more and more national trophy hunting organizations have exerted more and more influence on states' wildlife management. Increasingly financing seemingly good projects, more and more highly influential global trophy hunting organizations have started to infiltrate Montana's wildlife management. For example, the Governor appointing a member of the Safari Club International to Montana's Fish, Wildlife and Parks Commission would have been rather unthinkable prior to 2021.

These developments make a strong case against the NAM, showing that its approach is open-ended in terms of sportsmen's values, and that it is vulnerable to the pursuit of profit and special interests that trump science. The brutality involved in many policies shows an extremist view of certain wildlife species, and it has become clear that the organizations that follow the NAM fail to reign in the increasing number of extremists among their own ranks. Much of this is financially supported by the international trophy hunting and trapping industry.

Instead of managing ecosystems for the benefit of the ecosystem itself, the NAM has proven that it more so relies on species-specific hunting and trapping protocols and operates under management by numbers, and from a profit-seeking perspective when it comes to policy. Itself grounded in the exploitation of wild animals and in service to sportsmen's interests in hunting and trapping, it is grounded in mass exploitation instead of protection of wild animals. In this way the NAM serves a special interest group —that of sport hunters and the industries who profit from hunting—and it is perpetuated by that interest group to maintain exploitative values and influence policy to perpetuate such exploitation. Because of the NAM's reliance on individuals and organizations all profiteering from the exploitation of either wild or domestic animals, or both, for example, the corporate livestock industry, ranchers, trophy hunters, and trophy hunting organizations, sportsmen, trappers, large-scale fur 'operations, and the fur industry, the organizations who promote the NAM have opened their doors even wider to extreme commercialization and privatization interests.

State Wildlife Agencies: Outdated, Destructive, and Failing All of Us

As plant and animal species are going extinct, and many wild animals are losing ground and are forced to retreat further into their former habitat ranges at lower numbers, what they need most is broad, strong support by the public. What is needed most are changing our belief systems and a new model that promotes the connection and positive emotions, and promoting co-existence, the implementing non-lethal methods to resolve human-wildlife conflicts. None of this is of interest to the NAM and its promoters.

Instead, the NAM is guilty of promoting aggressive and harmful human behavior, and overall questionable behavior of those who enjoy the killing of wild creatures. This works against educating people about supporting wild animals and their habitat, non-lethal solutions for a peaceful co-existence with wild animals and appreciation for nature, but rather, it plays into the hands of those who want a quick, easy and lethal solution. As Rutberg has warned, "[M]any of the same people who push to invite bowhunters into a community to deal with 'nuisance deer' will later work to rezone open space for strip malls."[58] If anything, governmental promotion of lethal management in this context increases hostility against wild animals, including within the 'sportsmen' community especially with respect to predators, and indifference about their lives and homelands.

Promoting the killing of wild animals creates a host of adverse and undesirable reactions from the general public. First, a majority of the public is opposed to the killing of wild animals for reasons other than 'food' procurement, and an increasing part of the public opposes all killing of wild animals for any reasons and wishes non-lethal methods to be implemented, while yet others work towards a representation (proxy) of wild animals to give them a voice in the decision-making process—as of yet, none of their values are represented.

State wildlife agencies (SWAs) are financially dependent on those it supposedly regulates, and it is this dependency that makes these agencies favor the interests of hunters and trappers over those of the 'non-consumptive' public. Wild animals do not benefit from this system.

The myopic focus on a few game animal species distracts the public from the sheer unlimited types of human-caused harms that threaten the survival of wildlife species in the 21st century. This old system of wildlife management and conservation fails all of us because it is not promoting peaceful co-existence with true regard, respect, empathy, and compassion for wild animals, neither does it inspire awe and passion to protect individual wild animals, their mates, and ignores the urgent importance of wholesale protection of wild animals, their habitats and ecosystems on which we all depend on.

Thus, replacing the current business model of SWAs, the customers should be wild animals. Instead of management of wildlife, SWAs should manage *for* wild animals, along with a longer view that also takes the interests into account of future human and nonhuman animals. For the last 150 years, wild animals have been managed for human interests, now, and particularly under the threat of climate change and catastrophic levels of plant and animal extinctions and climate change, the focus of (motivation?) must change from 'consumption' and sportsmen's interests to protect and self-interests of wild animals. This, in my view, can only happen when the public no longer provides support for hunting and trapping, and instead, recognizes both for what it is: questionable behavior of a minority that enjoys killing wild animals for such frivolous reasons as recreation, trophies, and private profit.

Notes

1. Leopold, *Sand County Almanac*, 224–225.
2. Foran, *Subjugation*, 34.
3. American Game Protective Association, "Proposed."
4. Leopold and Krausman, "Science," 49.
5. Bolen and Robinson, *Wildlife Ecology*, 2.
6. Mills, *Conservation*.
7. Ibid., 4.
8. Ibid., 5.
9. Note: Their population size and survival rates; specifically, the estimated population size equals the ratio between the number of banded birds times the number of shot birds divided by the number of banded and shot birds.
10. White and Bishop, "Bountiful," 42.
11. Ibid., 45.
12. Mills, *Conservation*.
13. Bolen and Robinson, *Wildlife Ecology*, 184–185.
14. Wolfe and Chapman "Principles," 107, in: Novak, *Wild Furbearer*.

15. Bolen and Robinson, *Wildlife Ecology*."
16. Mills, Conservation.
17. Seamans, "Mourning Dove."
18. Darimont et al., "Unique Ecology."
19. Zeckhauser, "Human Hunters."
20. Lynn "Between Science and Ethics," 191, in: Lavigne (2006).
21. Spencer, *Gay*, 25.
22. Lynn, "Between," 197.
23. Spencer, *Gay*.
24. Ibid, 23–24.
25. As illustrated by Tom Krause, a former president of the National Trapper Association (NTA) puts it, "Our definition of 'Conservation' is 'wise-use'" (2008, p. 9).
26. NEFRTC, "Trapping and Furbearer," 12.
27. Holleman, "Wolf Populations."
28. "One mechanism for the formation of communities is homophily, or assortative mixing, the preferential association of individuals with others who are like them in some way (Newman 2002). In human societies, individuals have been observed to associate along lines of race, gender, age, income and nationality" among other factors (Lusseau and Newman 2004, p. 477).
29. Muth and Jamison, "Destiny of Deer Camps," 841.
30. Regan, *The Case for Animal Rights*, 354.
31. Kemmerer, "Hunting Tradition," 7.
32. "With the advent of advanced agricultural techniques, refrigeration, and extensive transportation, subsistence killing is obsolete for 99% of North American" (Luke quoted in: Kemmerer 2006, p.12).
33. Ibid., 8.
34. Van de Pitte, "Moral Basis," 259.
35. This story was conveyed to me in April 2019 by a former trapper, who wants to remain anonymous because he is afraid of harassment by trappers.
36. Taylor, *Respect*, 182.
37. Rollin, *Unheeded Cry*, 101.
38. Mahoney and Geist, *North American Model*, 137.
39. Corbett, "Reckless."
40. Brooks, 'Slaughtered Pups."
41. Macfarlene and Preso. "Wolf Killing."

42. Koehler, "Rocky Mountain."
43. RMEF, "Renews Support."
44. Ehrlick, "Believe Gianforte."
45. Associated Press, "Yellowstone Park wolves shot."
46. Wolfe et al., "Montana Wolf Policies."
47. Servheen et al., "Scientists."
48. Inman et al. "FWP Annual Report 2018."
49. Wilkinson, "Yellowstone Wolves."

50. In 2018, a total of 259 wolves were killed by hunters and trappers. An additional 60 wolves were killed in response to depredation—43 wolves died at the hand of US Wildlife Services, and another 17 wolves were killed by private citizens, likely landowners/ranchers, another 9 wolves were documented as illegally killed and 5 wolves died due to a collision with either vehicles or trains: Inman et al. "FWP Annual Report 2018."

51. Scott, "Hunters and Trappers."
52. Ashe, "Cruelty Toward Wolves."
53. Peterson, "25 Years After."
54. Wielgus and Peebles, "Effects of Wolf Mortality."
55. Center for Biological Diversity, "Wolf Experts."
56. Inman et al., FWP Annual Report 2020.
57. Lutey, "Brutal Winters."
58. Rutberg, "Why State Agencies Should not Advocate Hunting and Trapping," 34.

7

Crime Scenes in the Woods: The NAM and Cruelty Against Wild Animals

Laws as a Social Construct Protect Human Interests

As social constructs, laws and policies exist to regulate our treatment of animals, but that legislation is designed to protect human interests, rather than those of the animals themselves— with the NAM being a prime example. And although the law recognizes that animals merit some protection, the protection granted is contingent upon human assessment of their value. Animals we categorize as 'companions' are somewhat protected through anti-cruelty legislation, while animals we think of as 'food' or 'lab animals' are subject to broad exemptions that sanction nearly any violence against them, and wild animals are exempted altogether. Nonhuman animals hold the status of 'things,' as human property, to be exploited as replaceable commodities and subjected to commercial transactions—alive or dead, or their body parts. From local to global levels, the capitalization and legal exploitation of animals coincide with illegal acts against them. This occurs in manifold ways—as the poaching of wolves, but also the illegal trafficking of wild animals and their parts as the fourth-largest transnational crime in existence. Illegal animal

A. Heister, *Beyond the North American Model of Wildlife Conservation*, The Palgrave Macmillan Animal Ethics Series, https://doi.org/10.1007/978-3-031-14149-2_7

trafficking is in turn linked to organized crime, terrorism, as well as drug and human trafficking. Viewing wild animals as 'something' for human exploitation rather than as sentient subjects with intrinsic value, shapes wildlife 'protection.' Human-wildlife conflicts are regulated, and result in the victimization of wild animals.

Despite the last fifty years that have seen a dramatic rise in nature awareness and environmental consciousness, wild animals have not benefitted from this expanding circle of concern. Instead of protecting them as individuals who have the right to reside safely in their own communities, as members of *their* species and ecosystems, wild animals are harmed through laws and institutions, such as the organizations who follow and use the NAM which significantly harms and exposes nonhuman animals to human aggression and violence and endorses their injury and premature death in the name of "traditional" wildlife management and conservation. In theory, laws exist to protect wild animals along with companion animals and what is termed livestock. In practice, all laws and policies fall short of providing any group with actual legal rights or legal personhood.[1]

Animal Cruelty: PACT

When we talk about cruelty to animals, we mostly address cruel and violent acts against nonhuman animals under the control of humans (companion animals), but rarely consider traditional wildlife management as such. For example, on November 23, 2019, the much needed "Preventing Animal Cruelty Act and Torture" (PACT) Act was signed into law, which makes extreme acts of animal cruelty a federally prosecutable crime for the first time in the United States, punishable by severe fines and up to seven years in prison. While several activities such as impaling, drowning, and suffocating animals are now considered a federal crime against nonhuman mammals, birds, amphibians, and reptiles, similar methods are commonly practiced by sportsmen and continue to evade the public's attention entirely. For example, beating and stomping an animal in a trap to death, striking animals with a high-powered bow (this is impaling, is it not?), shooting a bird out of the sky, causing her/him to crash on the ground and then breaking that bird's neck; drowning a beaver caught in a Conibear trap set in water, are common methods trappers use to 'finish off' the helpless, exhausted, and suffering animal.

Wildlife management practices and lawful hunting and trapping are generally exempted from state-level anti-cruelty laws, which has the effect of further concealing the violent treatment of wild animals by sportsmen. Further, legalizing and promoting hunting and trapping by governmental agencies has the effect of hindering the public's awareness of hunting and trapping as acts of violence. If acceptable hunting and trapping methods were done to dogs or other companion animals, they would be considered abhorrent and unacceptable violent acts of animal cruelty.

Animal cruelty has been considered 'antisocial' behavior and involves aggression and violence against sentient beings.[2] The term "animal cruelty" is extremely narrowly defined and thus insufficient for this chapter's purpose in addressing hunting and trapping as acts of human-caused violence (as opposed to nonhuman violence[3]), which is at the root of the high levels of violence rampant in our Western societies. I find the definition provided by the World Health Organization (WHO) most helpful in exploring the issue of violence inherent in sportsmen's actions while at the same time representing the human-centered social construct that excludes violence against nonhuman animals despite the well-established link between animal abuse and violence against humans. For this chapter, I propose violence as "the intentional use of physical force or power, threatened or actual, against oneself, another person [or non-human animal—domestic and wild], or against a group or community, which either results in or has a high likelihood of resulting in injury, death, psychological harm, mal-development, or deprivation."[4] 'Intentional' is a pivotal part of what constitutes behavior deemed violent. I submit here that hunting and trapping are not only acts of intentional killing of other sentient beings, but furthermore, sportsmen seek intentional, violent encounters with wild animals that often result in injuring and otherwise harming wild animals, their mates, and families. As humans, we strive toward the protection of bodily integrity and consider physical and emotional harm as a violation of human dignity for good reason, whereas we see nothing wrong with hunting and trapping, both of which destroy animals' bodies and minds.

Cruelty Against Nonhuman Animals: Dimension

Defined as "behavior that causes unnecessary pain, suffering, distress, or death, animal cruelty can be physical or psychological and it must be demonstrated as deliberate."[5] Cruel treatment of animals affects billions of sentient animals globally, most are unrelated to human survival. This includes animals suffering on large-scale factory operations and killed in slaughterhouses, fur operations, circuses, zoos, rodeos, bullfighting, puppy mills, domestic violence, live animal transport and shipping, horse racing, and hunting, fishing, and trapping. This is cruelty socially accepted, legal and institutionalized, fully endorsed, and even promoted by respected institutions and democratically elected governments. And it is against this vast backdrop, explains Australian academic psychologist and animal protector, Eleonora Gullone, that socially *un*acceptable and illegal acts against animals are recognized as such.[6] In other words, when we talk about animal cruelty, we only focus on the comparably few and most egregious cruel acts against animals, while an enormous amount of human-caused animal suffering and deaths goes 'unrecognized.' This brings us to the main issue discussed in this chapter—hunting and trapping here understood as human violence and crimes,[7] similarly unrecognized as such.

Historically, concern for the welfare and rights of animals set in motion the welfare movement for children. New Yorker aristocrat and diplomat, Henry Bergh is celebrated as the founder of the American Society for the Prevention of Cruelty to Animals (ASPCA) in 1866 as well for initiating the child protection movement. Despite their shared roots, however, the two movements parted ways when in the 20th century, the government began to oversee child protection services, and private humane societies made animal welfare and animal rights their mission. It has been recognized today that this separation hindered the awareness of the correlation between animal cruelty and domestic violence.

Similarly, the exclusion of traditional wildlife management practices along with hunting and trapping from anti-cruelty laws and viewed as a 'separate' issue, is preventing us from recognizing hunting and trapping as cruelty against wild animals. This is even more of an obstacle given the constellation of who promotes and who protects wild animals, whereby today, it is the *government* that is even *promoting* killing wild animals for recreation and other frivolous reasons, while it is nonprofit organizations that are concerned for wild animals' welfare and rights. The difference is of great significance in terms of who is seen as the authority in these matters and who has more or all the power. This is to say that when it comes to wild animals and their treatment, it is the government that is creating laws, setting policies, and deciding what is legal and what is not (this has been discussed in Chap. 6).

Declaring hunting and trapping as 'legal,' has the effects of preventing us from recognizing that hunting and trapping involve animal cruelty, that they are activities reflective of destructive, aggressive, and violent human behavior, and finally, we may not take these issues seriously, after all, they are absolved by law. This is even further exacerbated by the choice of language describing hunting and trapping as necessary 'tools' for wildlife management. Given the lethal and devastating impacts on sentient beings affected, how is the callousness of the official language used here any different from describing rape as a "tool' of war or the shooting of innocent citizens as a 'tool' by the police to maintain law and order, all clearly unethical and unacceptable behavior?

Cruelty Against Nonhuman Animals and Interpersonal Violence: What We Know

The recognition that human-caused harm and crime not only affect humans but also nonhuman animals and nature has led to a 'greening' of criminology. In their 2021 publication called "Wildlife Criminology" British scholars of wildlife crime, Angus Nurse and Tanya Wyatt, introduce this new field that challenges the traditionally anthropocentric discipline of criminology. Viewing wildlife as the subject of 'criminological interests, the authors examine wild animals as victims in the context of commodification and exploitation, violence, rights and speciesism, and 'othering.' They recognize the interplay between the capitalization and legal exploitation of animals that coincide with illegal acts against them and interpersonal violence. Wildlife criminology aims to rectify the fact that, despite longtime advocacy for them, nonhuman animals continue to be kept outside of the discourses of human rights, and thus their interests and needs are marginalized in policy, legal protection, and social harm.

Research, in particular by American theologian and prominent author of animal rights issues Andrew Linzey has established a clear link between abuse of domestic nonhuman animals and violence directed against humans in domestic settings. For example, Linzey estimates that between 48% to 71% of battered women report that their companion animals have also been abused or even killed.[8] The progression thesis holds that offenders start by abusing small animals, then move on to abusing larger animals before escalating to violence against humans. However, as British criminology experts, Angus Nurse and Tanya Wyatt have stressed, not all nonhuman animal abuse automatically escalates into violent behavior targeting humans. They do note that nonhuman animal abuse and domestic abuse, particularly spousal abuse, are arguably linked as part of a continuum of abuse directed by male figures towards more vulnerable members of their households. The fact that adults who abuse nonhuman animals are likely to also abuse their children has led to the recognition of nonhuman animal abuse within families—in particular physical harm of the animal in the family—is now seen as an indicator for abuse of other family members to occur, typically perpetrated by the male in the family. In addition, nonhuman animal abuse has been identified as an indicator of psychological disorders with a potential propensity towards other forms of violence and antisocial behavior. In the context of wildlife criminology, Nurse and Wyatt contend "that this is also true of abuse and harm committed against wildlife."[9]

Relevant to this chapter is the research finding that treating most of the cruelty to animals as socially acceptable heightens both our tolerance for *causing* suffering to other species and our likelihood of causing it. Given such a relationship, the leniency with which animal cruelty is being treated by our laws, including in the judiciary, is a major cause for concern. Such leniency communicates the messages that trivialize individual animal lives and normalize aggressive and violent human behaviors directed toward other sentient beings. Consequently, only the most sinister and grotesque behaviors will sometimes attract maximum penalties.

However, even then, maximum penalties are trivial considering what empirical evidence indicates about the implications and correlates of animal cruelty.[10]

Animal Cruelty and Children

Children who have witnessed violence—either against humans or animals—are themselves likely to engage in abusive/violent behavior. Cruelty to nonhuman animals can be the result of a "compromised ability to experience feelings of empathy."[11] As a letter to the editor by a former trapper explains, the cruelty against pigs and chickens he witnessed as a child made him unaware of and indifferent to the cruelty he inflicted on wild animals.[12] As of 2016, the Federal Bureau of Investigation is required to track socially unacceptable animal cruelty alongside federal crimes including arson, rape, and assault.

Nurse and Wyatt refer to an analysis provided by Cohn and Linzey (2009) of pigeon shoots in Pennsylvania where schools are closed on the first day of hunting so that children can go hunting. The results of the study indicate that teaching children to kill nonhuman animals from an early age, can become part of their development that makes future violence more likely as well as an escalation into other forms of violence.[13]

Sportsmen, in particular men, indoctrinate their children into hunting and trapping at an early age. The importance of recruiting children at a young age into hunting and trapping to ensure the next generation of sportsmen and preserve their 'hunting tradition' has long been recognized by state wildlife agencies and sportsmen's organizations alike. State wildlife agencies along with the gun industry both depend for their survival on recruiting the next generation of shooters.[14] Thus, exposing young children to firearms and traps while acclimating them at an early age to the violence of killing wild animals is seen as crucial to occur before the child is capable of reflecting on the moral issues related to the killing of animals and before they make the choice to get involved in other activities. Consequently, there is an increasing trend in US states to lower the hunting and trapping minimum age, promote youth shooting programs among other strategies all geared to normalizing violence against wild animals.

A lawsuit seeking a temporary injunction against trapping in New York was supported by an affidavit provided by a psychologist specializing in the treatment of children, adolescents, and young adults. This expert pointed out the consequences of encouraging young children not only to commit acts of violence but even worse, to commit violence against helpless sentient beings who, when caught in traps, are not able to fight back. He further noted that in his experience, young people demonstrate compassion for animals and concern for their wellbeing, and these values are carried into their relationships with humans. Similarly, violence by young people also generally originates from the behavior and sanctions of adults. He concluded his statement with a warning that "children who are taught violence carry those patterns into their relationships with people and are psychologically primed to commit acts of destruction and violence upon others." Finally, he noted it is irresponsible for authority figures, particularly governmental authority figures, to add to the already existing distress young people are experiencing about violence in today's world.[15]

The danger of sanctioning animal abuse by children has been summarized by Margaret Mead (1964), who stated that, "The most dangerous thing that can happen to a child is to kill or torture an animal and get away with it." Childhood cruelty to animals as an "important predictor of later antisocial and aggressive acts and children showing these behaviors, without intervention, are at risk for enduring disorders in conduct and mental health." Studies have documented that exposure to violence, including violence to animals, can have long-term effects on the developing brain and subsequent poor health outcomes.[16] A 2013 report found that "23 school shooters reported that 10 (or 43%) had a background of animal cruelty, 90% of those incidents were 'up close and personal' attacks on animals, and 70% of the animals were unknown to the abuser (not a family pet)." Animal abuse, particularly aggression towards animals, was included in the Diagnostic and Statistical Manual of Mental Disorders (DSM-IIIR) in 1987 as a symptom of Conduct Disorder, which is defined as involving "repetitive and persistent patterns of behavior that violate the basic rights of others and societal norms or rules." Animal abuse has also been correlated to teenagers who are described as displaying callous and unemotional traits, which could be linked to deficits in empathy. These findings have resulted in suggestions for new policies for mandating cross-reporting of animal and child abuse between animal health care providers, animal control agencies, child health care providers, and child protection agencies.[17] However, state wildlife agencies are not included in this process.

Cruelty Against Wild Animals and Interpersonal Violence

As examined by the late Marti Kheel, hunting is an act of violence, and so is trapping. A study by US sociologist and expert of relations between animal abuse and interpersonal violence, Clifton Flynn examined the relationship between hunting and illegal violence among college males. Flynn concluded that hunters were about twice as likely to have been violent toward nonhuman animals; however, the study determined that one type of violence—killing wild or stray animals—accounted for this difference. Another, albeit limited conclusion nevertheless linked hunting related to harming nonhuman animals in the wild with property damage but not of nonhuman animal abuse or violence against humans. Flynn's study consisted of limited sample size, however, it should also be acknowledged that there has been little to no research done linking hunting and trapping with other types of violence, and this leaves the question unanswered whether there is a link between harming animals in the wild and violence against companion animals and against humans, and secondly, whether harming wild animals by sportsmen serves as a 'valve' for aggression and violent tendencies that otherwise would be directed against companion animals and humans.

Casual Acts of Cruelty

The NAM's legal cover for hunting and trapping conceals not only an overall gross violation of wild animals' rights to live unharmed by humans but also casual acts of cruelty. Sportsmen engage in killing unsuspecting and defenseless wild animals lured in front of rifles, guns, arrows, and into traps and snares. Deer and elk are often injured by arrows shot into their heads or bodies, or by gunshots, and suffer greatly at the hands of sportsmen. A Montana Outdoor/Hunting Guide that advertises elk trophy hunts, explains, "Elk often don't die as easily or quickly as deer, even with a killing shot. Sometimes several shots are required. Mortally wounded elk will often walk or run several hundred yards or more before dropping."[18] The high rate of injuries in wild animals has even been confirmed in government reports in different states, which all showed high wounding rates of deer, such as 51% in Montana and Texas, 58% in Michigan, and even 63% in Vermont. Finally, a study by the Ohio Department of Natural Resources found that after shot by a trophy hunter, on average, deer ran and stumbled 74 yards before finally dying. "The animal would be severely wounded and in severe pain and distress for a considerable period … In some cases, it took more than an hour for the hunter to finally locate the animal."[19]

Waterfowl do not fare any better at the hands of shooters; ducks are often lured onto the water by decoys, or they and other birds get shot while in flight. The latter is called "wing shooting" and is so popular that many businesses such as Trek International Safaris exist that advertise "wing shooting destinations" globally for hunters who want to combine travel, adventure, and shooting birds (primarily ducks and doves). How many of the 20 million doves shot by hunters every year in the United States alone, actually die immediately? This is unknown, but what is known is that after getting shot while in the air, a bird will crash onto the ground or into a body of water, will then get grabbed by either the hunter or his/her dog, and if still alive, may get her neck wrung or her head bitten by the hunter, "commander-style." This method is common among hunters 'finishing off' an injured duck. Duck/bird hunter online forums are illuminating other methods equally appalling, with storytelling about horrific treatments frequently joined by "laughing" emojis.[20]

Spring Bear Hunting

"Montana black bear are fair game in both the fall and spring!" states Swan Mountain Outfitter's website, announcing its services for guided bear hunts. Indeed, Montana is one of several states that continue to allow 'sport' hunting of bears in the spring *and* the fall of both males and females. Spring bear hunting season begins mid-April through the end of May, the archery-only season takes place in September, immediately followed by a 'general season' from mid-September through the end of November, during which a plethora of weapons are allowed, such as rifles, handguns, shotguns, and muzzleloaders, bow and arrow, and crossbows. In April 2021, Montana's Governor Greg Gianforte signed a bill into law that establishes an additional season that allows hunting black bears with dogs, and also a separate "hound training season" between the end of the spring bear hunt through mid-June, during which black bears can be harassed by dogs but not killed.

Despite state wildlife agencies honoring the NAM and its alleged 'fair chase' code, hunters shoot these bears at the time when they are most vulnerable. For example, after 4–5 months spent in the den without any food, black bears in Montana emerge from hibernation around the beginning of April weak and hungry. The high level of vulnerability is often exploited by outfitting companies and hunting guides advertising the benefits of shooting bears in the spring versus fall, whereby spring offers a near guarantee for hunters' success.[21] This has not only to do with the bears' own densities and available habitat but also because hunters know that after hibernation, bears will be preoccupied with searching for food (new grass on logging roads, blossoms, berries, and whatever else they can find) near their dens as opposed to bears' behavior in the fall when they travel and roam further.

Many females emerge with new cubs who are completely dependent on their mothers. Mother bears often leave their cubs in the safety of the den, or they hide them in trees in search of food to nourish themselves. Not surprisingly then, appearing in the hunter's scope without their cubs and as a single bear, many mother bears are shot, leaving their young behind who then may starve to death. In recognition of the cruelty of shooting female bears—potential mothers—and the likelihood of causing the subsequent starvation of their cubs, the public was successful in achieving a temporary suspension of spring bear hunts in the state of Washington for 2022. Previously, Washington was among the seven states that allow the killing of black bears in the spring (Arizona, Idaho, Montana, Oregon, Utah, Wyoming). Certain states such as Montana prohibit the killing of cubs, or females with cubs, and requires a mandatory check of all black bears killed.

Bear hunters are required to pass a one-time test to learn to distinguish between a grizzly and a black bear. No such test is required to learn how to distinguish males from female black bears because female bears *are* just part of the year's intentional targets. The sale of hunting licenses for bears is extremely lucrative for state wildlife agencies. As the 2020 Montana FWP 'harvest' report notes, the agency sold a total of 40,462 black bear licenses (at $37 for most residents) to sportsmen who then killed a total of 1598 bears, of whom 525 were females. Of the total of 1039 black bears killed in the spring, 342 were females; of the total of 559 black bears killed in the fall, 183 were females.[22] The state agency does not track the number of nursing bear mothers killed as part of the year's harvest, and neither is the number of cubs known.

Some mother bears and their cubs are not even safe in their den. For example, in April 2018, a man and his teenage son in Alaska were charged with illegal hunting after killing a female black bear in her den in southern Alaska and then shooting her two "shrieking" newborn cubs. The incident was caught on a camera that unbeknownst to the animal killers had been installed near the bears' den to observe them.[23] The consequences for the killers of the bear family? The son got a suspension and the father a three-month jail time. Defendants rarely get jail time in these types of cases because they claim that they kill the animal by accident.[24]

Wildlife-Killing Contests

Sportsmen organize 'wildlife killing contests, events where hundreds of native wild animals, including coyotes, bobcats, foxes, and other wild beings are easily wiped off the landscape over the course of a day or a weekend. These events are usually promoted as entertainment, involve teams competing on the highest number of animals killed, the weight or the sex of animals killed, particular species killed, or smallest or largest body or body part of the animal killed. These types of contests frequently involve betting and are followed by a party where contest prizes are awarded. The bodies of animals killed are usually dumped at the end of these events. Eight states have enacted bans of certain types of killing contests (Arizona, California, Colorado, Maryland, Massachusetts, New Mexico, Vermont, and Washington) with California being the only state that has a comprehensive ban on killing contests targeting game species, furbearers, and nongame mammals.[25] Despite increasing public outrage, these events are even growing in popularity nationwide, with local businesses often sponsoring them and the powerful livestock and ranching industry and other pro-sportsmen organizations contributing to them and publicly supporting and defending them.

Vehicle Killing and 'Coyote Whackin'

Wyoming and its neighbor, Montana, are two states where sentient wild animals can be deliberately killed year-round as part of 'thrill-kill" activities like the chasing and running over of exhausted wild animals with vehicles. Here, wild animals such as coyotes, are first being chased with recreational vehicles (Off Road Vehicles-ORV, or ATVs) and run over and plastered into the snow. If the animal is not quite dead yet, the sportsman may either shoot or slam the animal against his vehicle, an activity termed 'coyote whackin.' In 2018, then-Senator Mike Philips (Democrat) introduced two bills into the Montana legislature, which, if passed, would have ended wildlife killing contests, and the use of vehicles to chase and kill wild animals. The bills were treated with much disdain by most (white and male) legislators, who apparently failed to see the importance of the gross cruelty against wild animals the bills intended to end, and consequently, neither was passed. A 2018 online article fittingly titled, *A Death of Ethics: Is Hunting Destroying Itself?* by author and wildlife advocate, Todd Wilkinson, detailed these and other gruesome activities by sportsmen.[26] It speaks volumes that while the public has overwhelmingly supported my online petition to end this activity in Wyoming, none of the wildlife agencies or sportsmen's organizations have addressed, let alone condemned this practice.

Prairie Dog Shooting

Sportsmen also shoot wild animals that fall outside of the official hunting regime such as ground squirrels and prairie dogs. In Montana, these wild animals can be shot for recreation year-round, without a permit required on most lands. Considered an ecologically crucial 'keystone species' with a complex social structure and sophisticated language, prairie dogs are also considered shooting targets and "are often blown away by some to see the red mist of their blood hanging in the sky."[27] Ground squirrels and prairie dogs support 160–200 other animal species, and their burrows provide habitat for many other species, including burrowing owls, and black-footed ferrets, an extremely endangered species. Labeled by ranchers as 'vermin,' these wild beings, who "love to kiss" (see Jane Fonda's tribute of admiration to prairie dogs in Chap. 4), are not protected and can be killed by anyone at any time. For example, the Montana FWP classifies native prairie dogs as nongame wildlife, while the Montana Department of Agriculture classifies them as an agricultural pest (neither classification, as false as they are, provide any protection for prairie dogs). Thus, except for some wildlife refuges and a few other areas, most of the state's public lands are open for anyone to shoot prairie dogs, no license required, no questions asked. This carnage occurs even though Montana lists black-tailed prairie dogs as a 'species of concern', meaning the species is at risk due to declining population trends, threats to their habitats, restricted distribution, and other factors such as relentless persecution.

Poaching of Wolves: Officially Underestimated

Illegal acts against wild animals are committed by governmental officials, sportsmen and other private citizens. Predators such as wolves are especially vulnerable, but 'game' animals also become victims of these heinous crimes. As described in Chap. 6, Montana's Governor Greg Gianforte illegally killed a trapped Yellowstone wolf in February 2021 and received a warning, no penalty, no fine. A few weeks later, two young men shot two adult wolves from a helicopter, claiming they mistook the animals for coyotes. Neither had a wolf hunting license and shooting wolves from helicopters is illegal in Montana without permits, specified location, proof of need for livestock protection, and reporting requirements.[28] The state wildlife wardens did not want to know the name of the helicopter's pilot, even though pilots are as responsible for the illegal killing of wildlife from the air as the shooters, and simply charged each man a small fine while downplaying the severity of the incident—yet another example of a state wildlife agency shrugging off illegal acts against wild animals.

In a mass shooting event in the fall of 2020, about 100 hunters surrounded a herd of elk, and opened fire on the animals, shooting from the road and both sides of the area, injuring many and fatally killing 50 elk. Montana's state wildlife agency downplayed this aggressive and violent event and declared it unethical but not illegal, citing six of the hunters for minor violations, e.g., having had an outdated hunting license.[29] The poaching of wolves is a serious, yet unaddressed problem under the NAM. The February 2021 poisoning of eight wolves in a county in Oregon[30] is just one of many illegal acts against wild animals.

A study review found that the poaching of wolves has systematically been underestimated by biologists and policymakers and that "[T]he *underestimates* have obscured the magnitude of *poaching* as the major threat to endangered wolf populations" in the United States.[31] Furthermore, others have observed that most wolf-poaching incidents remain unresolved and they estimate that for every poached wolf found, one to two additional illegally killed wolves go undiscovered.[32] These heinous acts against wolves and other animals are ongoing under the NAM. How are these killings of highly sentient and social wild animals any different from hunters (legally) shooting and trapping 20 wolves each as soon as they step across the park's protective boundaries?

Spotlight on Trapping: Torture Before Death

"Few men could endure to watch for five minutes an animal struggling in a trap with a crushed or torn limb," wrote Charles Darwin in his essay "Trapping Agony" in 1863. Darwin went on to express his utter astonishment that such animal cruelty was permitted to continue in his modern times. He felt that the acts of cruelty perpetrated by trapping were inhumane and morally wrong. His conviction was that cruelty carries more weight than the claim that setting steel traps is the only way to preserve game.[33]

The reality in the 21st Century is that nearly 2.8 million wild animals died in traps and snares in 2018, down from nearly seven million wild animals were reportedly killed by recreational and commercial trappers nationwide in 2011–12.[34] Other reports estimate that between six and 21 million wild fur-bearing animals are killed by trappers annually in the US.[35]

No matter how staggering the statistics, numbers alone cannot convey the tragic circumstances under which animals suffer in traps and snares prior to their forced, premature death, nor do they capture the depth of the violence experienced by trapped animals at the hand of trappers leading to their deaths. Each year, in an extremely drawn-out ordeal, wild animals are caught in baited and concealed traps and snares, in whose grips they die either immediately if they are 'lucky', slowly, agonizingly, and on their own, or must linger in a perpetual state of bodily injury, panic, pain, and fear, until the trapper returns to kill them. Each of these millions of animals—including so-called non-target animals who were not the animal whose fur the trapper wanted (trappers refer to them as "trash" animals), and members belonging to an either endangered or threatened species such as Canada lynx—was first lured into a concealed and baited body-gripping trap (foothold trap or snare catching a limb), or into a spine-crushing device (Conibear trap), or caught by a neck snare. If animals are still alive when the trapper returns, perhaps after barely surviving for several days and nights while exposed to extreme, freezing temperatures and attacks by other predators, trappers prefer to either stomp or beat them to death (often using metal or wooden clubs), shoot, drown, or strangle the animal, choosing methods least harmful to their pelts. Sewell Newhouse (1865), who mass-produced the first commercial leghold trap in 1852, recommended against the use of poison such as strychnine, or shooting trapped animals; he considered these wasteful practices because of the risk of damaging the fur, which subsequently reduces the trapper's profit.[36]

Types (and Dangers) of Common Traps

The steel-jaw leghold trap, banned in over 80 countries and condemned as inhumane by the American Veterinary Medical Association, the American Animal Hospital Association, the World Veterinary Association and the National Animal Control Association, is still the most commonly used trap in the United States today. Apart from some modifications, for example some states no longer allow toothed traps, most leghold traps now leave a tiny gap so as not to ruin the fur, and to deflect critics, rubber padded traps are available, the trap's design has essentially not changed since Sewell Newhouse invented it back in 1823. Scientific trap research has shown that the results of 28 out of 38 studies on leghold traps were below currently accepted animal welfare standards, and that leghold traps, being extremely stressful for the animals, can have "poor capture specificity and reduce survivorship of released animals."[37] Furthermore, injuries caused by leghold traps include severe swelling, lacerations, dislocated joints, fractures, damage to teeth and gums, limb amputation and death.[38] Self-amputation is common enough that trappers have a name for it: *wring off*. If these amputees don't die from blood loss, they are likely to perish from starvation. Animals caught in steel-jawed traps (as with snares) also die from exertion, starvation, or exposure in cold winter temperatures, while others perish when predators happen by and attack. In addition to physical injuries, Proulx has shown that the psychological trauma of being caught in a trap can alter the behavior of released animals and reduce survival rates. Other impacts caused by psychological trauma, such as fear and anxiety, or physiological stress and pain have been assessed in trapped animals through behavioral analyses and stress-related hormonal and blood-cell measurements. Furthermore, findings show that trapped animals have increased levels of serum cortisol, which reflects their stress response to being caught, while other studies show an increased heart rate and body temperature during the initial period of capture (Fox and Papouchis 2004).

The Conibear—a body-gripping or killing trap—is commonly used and set underwater for semi-aquatic species, such as muskrats, mink, beaver, and river otter. For a Conibear to instantly cause death by a killing blow to the neck or head, several variables are needed— the correct size of the animal entering the trap at the correct angle and speed.[39] For these species who are physiologically adapted to aquatic life (for example, otter can dive up to 22 minutes, muskrat for 12–17 minutes and beaver for up to 15 minutes) "death by drowning-induced hypoxia is a slow process." In these animals, while struggling to reach the surface, which consumes oxygen more quickly, loss of consciousness occurs after an average of four minutes for minks and nine minutes for muskrats, and 10–13 minutes for beavers.[40] Since these animals struggle for more than three minutes before they die, researchers have concluded that traps set in water do not meet basic animal welfare standards and as such, are inhumane.[41] Moreover, and as previously discussed, death by drowning—induced hypoxia—is not considered an acceptable method of euthanasia by veterinary and laboratory researchers and does not meet the presently accepted standards for traps designed to kill animals.[42]

Animals caught in neck snares die of asphyxiation when the wire tightens as the animal tries to pull free. Although there is no research determining welfare standards of animals dying in snares, in studies where animals were released after capture, researchers observed that damage caused by the pressure of neck snares on tissue may take days to appear and animals frequently died as a result of the tissue necrosis.[43] Studies on carnivores have shown that comparably minor injuries such as broken teeth or claw loss can cause the inability to catch wild prey and lead to an increase in livestock predation.[44]

This brief overview of the scientific literature on trap research shows that restraining traps and killing traps commonly used by trappers do not meet internationally accepted standards of animal welfare, which are already low and incomplete. They also do not address the methods trappers employ to kill wild animals in traps and snares. As often advocated in trappers' magazines, these include suffocation, drowning, gassing, and hitting with clubs to minimize any damage to the animals' pelt. The Trapper Education Manual (2005) recommends the following method for inexperienced trappers who are encouraged to "use submersion techniques or kill-type traps… This avoids the need to kill your catch." For those trappers who are underage or otherwise not legally permitted to carry a firearm, the publication recommends they "strike smaller furbearers such as raccoon, opossum, and fox hard at the base of the skull with a heavy wooden or metal tool to kill or render them unconscious. Placing your foot over the head and chest area and compressing these organs will lead to death."[45]

The vast majority of traps currently in use were developed by trappers and/or the fur industry, thus the welfare of the trapped animal reflects the desire to obtain undamaged pelts, with the welfare of trapped animals being a secondary issue or deemed irrelevant.

The Carnage of 'Regulated' Trapping of Wild Animals

Montana

Between 2013 and 2018, trappers in Montana reportedly killed a total of 272,420 wild animals, including 26,323 beavers, 1734 gray wolves, 568 otters, and nearly 103,000 coyotes. For the 2013–14 season alone FWP sold nearly 6000 trapping licenses, which resulted in over 53,000 wild animals killed in traps and snares that season. However, as recent agency responses to citizens' Freedom of Information Act (FOIA) requests have shown, the carnage does not end here but also involves high numbers of so-called 'non-target species,' animals trappers did not intend to trap.

Between 2012 and 2017, trappers reported 349 non-target animals caught in traps or snares. Trappers are only required to report the capture of lynx, domestic dogs, and 'protected' animals (game and furbearer species). Of these incidents, the species the set was intended for was known in 251 cases. 55 percent of these animals were released, while 45 percent of them died. Of the 148 dogs caught, several were injured and 7 died. Mountain lions were the most common species incidentally caught with 67 individuals dying while 32 were released (17 percent of them were released with injuries); snares caused the highest mortality among mountain lions; all mountain lions caught in a snare died (54 percent, more than 34 individuals). The report's summary notes that "the rate of these wildlife deaths caused by incidental trapping is not a population-level concern for any species in Montana."[46]

Idaho

A FOIA request to the Idaho Department of Fish and Game (IDFG) revealed that between the 2012–13 and 2018–19 trapping season, trappers targeting wolves killed 813 wolves and caught 620 non-target animals, including rare fishers, wolverine, eagle, and lynx. Of these, 269 were released alive and 351 were killed.[47] We need to keep in mind though, that a percentage of those animals who were released alive later died of injuries sustained from the traps that either killed them or made it impossible to hunt or scavenge. A previous FOIA request submitted to IDFG by the nonprofit Western Watersheds Project showed similarly horrific and widespread capture and mortality of non-target species. During the 2011–12 trapping season alone, wolf trappers also found 246 non-target animals caught in their traps and snares, of which 116 were released alive, 118 were killed, and the fate of 12 other wild animals was not reported. Among these victims were 22 fishers—a furbearing species in the Northern Rockies under consideration for federal protection—of whom 15 died in traps or snares.[48]

Failed Attempts to Ban Leghold Traps and Define Humane International Standards for Trapping

An international debate over the cruelty inherent in trapping led the European Union (EU) in 1991 to adopt a precedent-setting international regulation aimed at prohibiting imports of fur coming from countries that had not banned leghold traps. This regulation (#3254/91) resulted in the EU prohibiting the use of leghold traps in all its 15 member countries by 1995. It failed, however, to also prohibit, after 1996, the import of pelts from a certain number of animal species from countries that do not ban using leghold traps. Because the U.S and Canada did not want their fur trade curtailed, both countries responded with a trade threat that would punish the EU economically under the General Agreement on Tariffs and Trade (GATT). Applying this pressure and side-stepping a potential ban on leghold traps and a fur export into the EU, the U.S. and Canada touted an initiative to develop Best Management Practices (BMPs) for traps and trapping methods that would cover 29 animal species, established a National Trap Testing Program (NTTP) and by 1997, had successfully averted the European threat of a fur ban.

More than $40 million have been invested to fund this massive trap testing program since its start in 1997. Agencies and organizations with vested interests in making sure that trapping continues were involved in providing support, development, research, and funding, such as the National Trappers Association, Canada's Fur Institute, the International Fur Trade Federation, as well as the U.S. Department of Agriculture (USDA) and state wildlife agencies.

The BMPs remain a mere voluntary program for the state wildlife agencies to implement, while it is a powerful strategy for them to promote traps and trapping nationally and internationally. Not only does the agreement fall short in providing basic welfare guarantees for all trapped animals, but it enables the trapping and fur industry to hide their abhorrent treatment of animals, and to continue their horrific practices and treatment of animals on large-scale fur 'farms' as well as catching wild animals in traps and snares, now aided by soundbites of an alleged commitment to 'humaneness' and 'scientific research, touting trapping as 'state-of-the-art conservation tool' while changing very little about the actual practices.

The BMP program presupposes the trapping of wild animals as valuable and that there is a humane method of trapping, and this presupposition undergirds the entire program. This is no surprise, as they were written by the AFWA's Furbearer Conservation Technical Work Group, which is composed of the agencies' own wildlife biologists. Among the declared goals of the BMPs programs is to promote (regulated) trapping as a modern wildlife management tool and to instill public confidence in and maintain public support for both. To this end, governmental officials are trained to better communicate trapping as part of the NAM, and how to discuss the importance and (alleged) necessity of trapping for wildlife management. The curriculum for their training includes the education on the BMPs along with profiles and motivations of the modern trappers. These governmental-sponsored, trapping-promoting workshops use wildlife biologists with the US Department of Agriculture/Wildlife 'Services' as instructors and are supported by a partnership between the AFWA, the USFWS, and an influential nonprofit, The Max McGraw Wildlife Foundation, whose goal is to secure a future for hunting, fishing and land management through programs of science, education, demonstration, and communication.

The manipulation of the public can be seen on AFWA's website, which presents the impressively long list of BMPs per 'species' (here is the species-level again to imply that all animals belonging to the same species are alike, e.g., all foxes are the same). Their website also makes various brochures and other documents available with talking points for trappers as well as communication strategies for sportsmen about how to talk with the public about the importance of 'regulated' trapping. BMPs do not address that animals' vulnerability is also exacerbated by the fact that traps and snares are generally baited often with parts of animal carcasses and thus indiscriminate. Neither do they specify the length of time that animals have to endure in these devices at high levels of suffering. Finally, the BMPs do not require a mandatory trap check time limit for trappers, which would at least reduce the length of time of animals suffering in these devices.

What Level of Animal Suffering Is Acceptable?

According to the Agreement on International Humane Standards (AIHTS), a restraining trap meets the standard of humaneness if 20 percent or fewer of animals of the same species show any signs of "poor welfare," defined as "self-mutilation, excessive immobility, fractures, severance of tendon or ligament, severe external hemorrhage, internal hemorrhage, skeletal muscle degeneration, spinal cord injury, severe internal organ damage, amputation and death." Apart from being impossible to enforce, this means it is acceptable that out of 90,000 coyotes, who were caught in leghold traps, 18,000 of these nonhuman wild animals can suffer these horrendous injuries and this would be called 'humane.' In terms of so-called "killing traps" (Conibear traps and drowning sets), it is acceptable for 20 percent of animals of the same species to endure a time period that exceeds 5 minutes between caught in a trap and until loss of consciousness for most species, except for 45 seconds for ermines and 2 minutes for martens. Apart from being entirely unenforceable (trappers are rarely present when an animal gets caught in their trap or snare), this is an extensive length of time for an animal capable of experiencing pain and panic to suffer, especially considering that beavers and other semi-aquatic animals often struggle violently for up to 20 minutes before drowning.

Furthermore, killing neck snares, which are among the killing devices that should render a trapped animal irreversibly unconscious within 5 minutes, are not included in the AIHTS even though these devices cause the same level of horrific injuries and suffering to wild animals as those caused by steel-jaw leghold traps (restraining traps). According to researchers, when canids are snared around their abdomen, this can lead to disembowelment. Furthermore, while edema is considered a minor injury in leg-captured canids, it is major in neck-captured animals who do not die quickly. Scientific evidence refutes trappers' claim that snares are humane and cause quick unconsciousness in trapped animals. In addition, researchers have critically pointed out that trappers do not explain why wolves, coyotes, and red foxes "are commonly found alive in killing neck snares with broken teeth and bleeding gums, swollen heads (so common trappers call them jelly heads), and lacerated necks … Capture sites often show signs of struggle, including blood, urine, and defecation …." They stated, "The fact is that unlike researchers … trappers do not sit and observe animals being caught in snares."[49]

Gilbert Proulx, a Canadian wildlife biologist with over 40 years of field experience and trap research came to the same conclusion. His 2018 book titled, *Intolerable Cruelty: The Truth Behind Killing Neck Snares and Strychnine*, describes the incomprehensible cruelty involved in the use of snares: trail cameras hidden near two snares set according to the standard outlined by the industry. The coyote was caught in the snare at 11:51 am. For the next 14 hours, 16 minutes and 14 seconds, the animal is conscious and violently struggling against the snare that tightens the more the coyote struggles. Proulx writes, "Struggling was intense and frequent at the beginning of the capture period and consisted of strong jumps, twists, falls, and attempts to bite the snare cable." The coyote struggled until the end. It eventually collapsed, and it lost consciousness 18 minutes later. I estimated time of death at 20:31:05, i.e., 23 minutes 51 seconds after irreversible loss of consciousness At time of death, the position of the animal suggested a quick and humane kill. However, the neck snare did not quickly render the coyote irreversibly unconscious. It was not a humane kill, based on existing humane criteria" (pp 10–11). The second trap victim, a wolf caught in the neck snare did not fare any better.

Proulx, who is a scientist and certainly not a "bleeding heart" but has decades-long experience of seeing wild animals in traps, writes, "It is with great difficulty that I watched these video recordings. I could not believe that trappers could do this to sentient animals. Simply put, these 2 animals had been tortured."[50]

Trapping of Beavers: The Case of Drowning

The Trapper Education Manual (IAFWA 2005) recommends drowning sets and claims that "semi-aquatic furbearers (muskrats, mink, river otters, beaver, and nutria) cannot take water into the lungs so, technically, they asphyxiate," implying that because the animals may not inhale water, death comes from CO2-induced narcosis and is thus painless. This is called dry drowning. However, studies on the near-fatal drowning experience of humans have shown that approximately 15% of them experience a dry drowning in which the lungs remain largely water-free. And here, observations have shown that a longer period of consciousness may occur in dry drowning compared to wet drowning, in other words, research suggests that death occurs more rapidly when water *is* inhaled.

Following their review of many studies that have intentionally drowned animals such as beavers, mink, as well as dogs, and rats among many other animal victims of research, Ludders et al. (1999) published their findings that reflect their conclusion in an article titled "Drowning is Not Euthanasia." They critically noted that drowning as a method of killing wild animals may be convenient and a traditional wildlife management technique, but the concept of euthanasia is independent of traditions and convenience. They concluded that because time is an important element in euthanasia, any technique that requires minutes rather than seconds to produce death cannot be considered euthanasia. Thus, drowning is not a 'good death' that occurs without pain and distress. The authors recommended that their findings be considered in developing guidelines and Best Management Practices for animal treatment, which, as far I know, has not happened, not even 22 years after the recommendation.

Trapping of Bobcats: The Case of Strangulation

To "strangle" someone means to obstruct seriously or fatally the normal breathing of a person, while "choking" means having the trachea blocked entirely or partly by some foreign object like food."[51] This means that how animals die when caught in a neck snare or a catchpole, resembles painful strangulation, not choking, a term frequently used by trappers. When caught in a snare or leghold trap, bobcats are known to engage in an aggressive and desperate struggle for their lives. An online video[52] of the rescue of a male bobcat caught in an illegally set snare shows the struggle of the bobcat and frantic movements back and forth, alternating between retreating as far as the snare that was wrapped around his body allows, and movements to attack while growling. Recommended methods for killing trapped bobcats include strangulation via a catchpole, shooting the animals in the brain (to prevent fur damage), or administering a sharp blow to the top of the skull with a metal trap stake or suitable club.

The same methods of strangulation and blunt force trauma are used in domestic violent and abusive relationships where the victims are predominantly women. Nearly all victims of strangulation are strangled manually (with hands) and almost all, or 97 percent of strangulation attempts, also involve blunt force trauma, notes the Training Institute on Strangulation Prevention website.[53] Strangulation has been identified as one of the most lethal forms of domestic violence, whereby unconsciousness may occur within seconds and death within minutes. When perpetrators use strangulation to silence their victims, they do this as a form of power and control. Trappers likely feel very much in control and extremely powerful when they end an animal's life, as in "a job well done." No research exists that I am aware of has investigated the emotions involved in how trappers experience their activity.

Strangulation of wild animals such as bobcats resembles mostly two of the four types of strangulation known in humans: Hanging and ligature (also known as garroting)—use of a cord-like object, clothing, rope, or belt. "When being strangled, the human victim will first experience severe pain, followed by loss of consciousness, then brain death."[54] It is not far-fetched to assume that this is the same or similar pain an animal experiences when strangled to death in a neck snare.

Intersectionality of Violence

The World Health Organization (WHO) views violence as "the result of the complex interplay of individual, relationship, social, cultural and environmental factors," and treats violence as a public health issue of global dimension. Today we know that violence does not occur in isolation but rather, animal cruelty and violence against humans, and crimes are linked. From studies on the workings of violent, oppressive systems, we have learned that they create, reinforce, and expand each other because they have the same root—a mindset that reinforces a socially constructed status hierarchy—regardless of who the violence is perpetrated against—animals, humans, or environment/nature.

This fabricated hierarchy is then knowingly and unconsciously applied to justify oppression and affects all aspects of our social and political life, including those institutions that create laws and social policy, which in turn, allow some groups to be privileged and prosper while others are victimized. American civil rights activist and attorney, Kimberlé Crenshaw coined the term 'intersectionality' in 1989, which describes the study of the overlapping, intersecting social identities and relating systems of oppression, domination, and discrimination. Intersectionality holds that classical conceptualization of oppression within society, including racism, sexism, homophobia, bigotry, and speciesism, do not act independently from each other. Instead, these forms of oppression interrelate and create a system of oppression that reflects the 'intersection' of multiple forms of discrimination (recall from Chap. 2 that Melanie Joy has called this 'powerarchy," an overarching system of discrimination). These conceptualizations are socially constructed to categorize sentient beings and then used to either value or devalue each other.

The WHO has developed an ecological model to try to understand the multi-faceted nature of violence, which assists in examining factors that influence behavior, or which increase the risk of committing or being a victim of violence. This framework begins by identifying the behaviors of the individual; involves personal relationships; explores the community aspect; and finally looks at the broad societal factors that help create a climate in which violence is encouraged or inhibited, including certain social norms. The summary of their 2014 Global Status Report on Violence lays out how widespread global violence is, and states that "Women, children and elderly people bear the brunt of nonfatal physical, sexual and psychological abuse; a quarter of all adults report having been physically abused as children, and that one in five women reports having been sexually abused as a child."[55]

The harming and killing of unsuspecting, unoffending, and defenseless animals should be treated seriously and as issues of immediate societal concern. Hunting and trapping do not just constitute individuals' choices that can be rationalized away by explaining that it benefits one's recreation, economic profit, tradition, lifestyle, or mental health. Regardless of who is the recipient of violence—women, children, domestic and wild animals—we know today that violence does not occur in isolation and may affect us all. From individuals, families to the community, through all strata of society, the disregard for individual sentient lives and the inherent lack or absence of empathy for 'others'—the moral glue that holds human societies together—violence should be of great social concern, particularly in the era of unprecedented environmental destruction, accelerating plant and animal extinctions and climate change.

How the General Public Is Affected

The violence of hunting and trapping harms and kills animals, and it impacts humans as well. In the past and ongoing, sportsmen have had a significant influence in writing hunting and trapping regulations and deciding what's acceptable. For example, with respect to Montana, trappers have resisted any changes or compromises brought forward by the public. For example, the public has requested that trappers post warning signs in areas they trap to help keep companion dogs safe. The public has also asked that a mandatory trap-check be implemented but trappers have objected rather successfully to any restrictions of their activities and done just the opposite. Through their influence on the Montana legislature, they have fought to expand trapping without giving an inch to the public concerned about the cruelty inherent in trapping, worried about their own, their children's and their companion dogs' safety when enjoying their own, tax-funded public lands.

This has led to considerable tension between the public and the trapping community.

The public and companion animals are exposed to the threat of encountering traps in the fall, winter, and spring months. Many people no longer use their own public lands out of fear and concern for the safety of their dogs. The fact that every winter many dogs are injured—some so severely they require amputation of a limb or sustain lifelong traumatization—shows that this concern is legitimate. In addition, many dogs die because of accidentally stepping in a trap or a snare that is usually baited. Montana's Fish and Wildlife Commission verbally promised to address a proposal written by Footloose Montana and submitted with four other conservation groups on July 28, 2020, to set aside high recreation areas on public land from trapping. The Commission postponed consideration for over a year and finally dropped it from the agenda.

For those who have a well-developed sense of empathy, knowledge about cruelty and violence against animals, and their prolonged trauma, often-severe injuries, and brutal modes of death, is an issue of mental and spiritual distress, thus, detrimental to mental health. In December 2013, a man in Oregon stepped in a conibear trap set for beavers. The trap broke his leg. Some people have either seen animals in traps or heard them scream in pain and fear before they die, both of which leaves scars on their mind and spirit. Many people who just wanted to enjoy the day with their dog outdoors have returned home with their beloved companion dead, having been killed in a trap. These people are often traumatized for years.

Others are luckier when their dog is "just" injured in traps or snares, but these injuries often result in amputation of their dog's toes, foot, or leg, which greatly impacts the quality of their lives. In addition, veterinarian bills for these incidents can be very costly. To add insult to injury, many people destroy the traps that harmed their dogs, an illegal act that makes them vulnerable to criminal charges. This leads to general paranoia about being in danger on public lands as well as the very real fear that the trappers will seek retribution by destroying property in retaliation and further harming their companion animals and livestock. As a result, many do not report the trapping incidents, and the public remains unaware of the real number of deaths and injuries of companion animals.

Frequently, people admit, either with a sense of defeat or anger that they fear for the safety for their dogs and no longer enjoy the vastness of Montana's public lands in the winter, or they only go to (urban) places where they know they'll be safe. This decision has led to some people abandoning their walks with friends because they are fearful of hiking in national forests in the winter months due to traps, while others may decide to join their friends but leave their dogs behind at home sp as to not endanger them. Either way, the world of the non-trapping public with companion dogs becomes very small during each winter's trapping season. Finally, the violence inherent in hunting and trapping also affects people involved in rehabbing wild animals. When injured animals are found and brought to them, wildlife rehabbers often invest a significant amount of their time, energy and money in the animals' care to help them regain their health. When animals are released back into the wild, their human caretakers know that they face the potential fate of injury, panic, pain and death in a human-dominated world.

Applying formal and statistical research in political science and economics, one may view trapping as a form of social polarization that potentially increases the likelihood of general violent conflict. Esteban and Ray (1994) have stated that polarization within society can result from the "interaction of within-group identity and across-group alienation."[56] Applied to the contemporary debate over sport hunting and trapping, this may mean that the rift between governmental agencies promoting hunting and trapping and supporting sportsmen, and the growing sector of the public opposing both, continues to deepen. This process could eventually lead to even more and violent conflict because both sides feel ideologically separated from members of the other group. As confirmed by research, ideological polarization can be "a major source of conflict and hence one of the key impediments to social and political progress."

Conclusion

Wild animals are harmed by human-caused violence in a multitude of ways, indirectly through ongoing species extinctions, and directly through hunting and trapping. Wild animals are competent, independent, sentient, and socially skilled agents of their own world, but sportsmen turn them into victims of human aggression, violence, and cruelty causing them injuries and premature death. Given the efforts of conservation to curb society's destructive relation to the rest of nature, it is disturbing to see the widespread acceptance and promotion of the NAM among conservationists, as it legitimizes another strand of destruction—violence against wild animals in the name of recreation, tradition, and management. The analysis above shows that sportsmen's activity—hunting and trapping—involve an unacceptable absence of respect and empathy for the lives of *individual* wild animals while allowing an incomprehensible level of violence and cruelty against them. NAM proponents' enjoyment of killing animals for pleasure is ethically unacceptable behavior and detrimental to society's efforts to increase pro-social behavior. On a pragmatic level, these activities are accelerating the destruction of the natural balance that all beings depend upon for survival. Considering the interlinked and worsening crises that are threatening all life forms on this planet, hunting and trapping can no longer be seen in isolation but rather as part of a multitude of human-caused violence and harm against wild animals. The violence inherent in hunting and trapping, and its root—the hierarchical mindset or relevantly the human supremacy mindset—should be of great concern to society, regardless of to whom it is directed, humans, wild and domestic animals, or nature.

Notes

1. Wise, "Legal Personhood."
2. Gullone, "Conceptualising."
3. Nurse and Wyatt (2021) write about how violence observed in nonhumans is different from human violence, and generally occurs as part of the struggle for resources, or in response to first being victimized by humans.
4. Krug, "World Report," 5.
5. Gullone, "Minimize Animal Suffering," 1.
6. Ibid.
7. I differ here from Nurse and Wyatt in that I disregard the anthropocentric differentiation between legal and illegal acts, the latter usually called 'crime.' If we consider the animal standpoint, this distinction is most likely irrelevant.
8. Linzey, *The Link*.
9. Nurse and Wyatt, *Wildlife Criminology*, 85.
10. Gullone, "Minimize Animal Suffering."
11. Nurse and Wyatt, *Wildlife Criminology*, 86.
12. Letter to the editor:

    "<u>Trapping: Steel traps should never be allowed</u>

    I received very little money from my parents while growing up and used steel traps to catch animals to earn a little money, never dwelling on the horrible pain and suffering the traps did to animals caught in them. I'd get up at 4 or 5 a.m. to run the trap line before school; any animal still alive was either drowned or finished off with a blow to the head.

    Since I saw farm animals slaughtered every day, I was indifferent to their suffering and deaths. A butcher knife dispatched chickens for dinner. Friendly ducks and geese who had trusted me and followed me about the farmyard eventually met the same end.

    Yet, somehow there was a disconnect between me, my pets and the animals I killed, as though they hadn't had the emotions, the feelings or thoughts I had. I was never admonished to be kind or considerate to animals.

    It's been 65 years since the morning I found a broad-striped skunk caught by my trap. It had snapped shut on her last good paw. She had been hobbling around on three stumps, all nicely healed, if one can describe them so, and her sole forepaw. Animals, you see, will chew or twist their paws off to escape. I had lost many traps by animals who desperately pulled trap stakes to escape. Some I later found dead, some hanging by their entangled traps and chain high up in a tree where they had died.

The cold indifference trappers still display today, as do the so-called "sportsmen" who hunt these creatures of God is deplorable.

I've regretted to this day having been so cruel - having tormented and destroyed those animals so long ago. But I can never undo that. Not ever.

We shouldn't allow steel traps anywhere in the wild" (Leonard Stastny).

13. Nurse and Wyatt, *Wildlife Criminology*, 85.
14. McIntire, "Guns."
15. Animal Legal Defense Fund, "Children and Trapping."
16. Phillips, "The Link," v.
17. Ibid.
18. file:///C:/Users/loxod/OneDrive/Book/Resource%20Material/ElkHuntingGuide%202014.pdf
19. Gonçalves, Exposed, 130.
20. My Screenshot: 2020-12-26.png - OneDrive (live.com);
    2020-12-26 (3).png - OneDrive (live.com);
    2020-12-26 (5).png - OneDrive (live.com).
21. Total Outdoor Adventures, "Spring vs. Fall Black Bear Hunts."
22. MDFWP, "2020 Bear Harvest."
23. Bever, "A Mother Bear."
24. Jacobs, "Alaska Hunter."
25. Animal Welfare Institute, "Wildlife Killing Contests."
26. Wilkinson, "Death of Ethics."
27. Wuerthner, "North American Model."
28. Montana Department of Livestock. "Aerial."
29. Associated Press, "People Cited."
30. "Eight Wolves Poisoned," The Guardian.
31. Treves et al., "Mismeasured Mortality," 1256.
32. Center for Biological Diversity, "Reward."
33. Darwin, "Vermin and Traps."
34. AFWA, "Fur Harvest."
35. (White et al., 2010, p. 68).
36. Newhouse, *Trapper's Guide.*

Note: In his promotion of the steel-traps, Newhouse points out its many benefits such as "…securing the animal caught from destruction by other animals, and from escape by self-amputation; and above all, the steel-trap *does no injury to the fur*" (p. 15). Newhouse considered shooting a trapped animal, "a very wasteful method" because "[F]ur-dealers and manufactures consider skins that have been shot (sic), especially by the fowling-piece, as hardly worth working. The holes that are made in the skin, whether by shot or bullets, are but a small part of the damage done to it" (Newhouse, p. 14).

37. Iossa et al., "Mammal Trapping," 345.
38. Proulx, "Review."
39. Gilbert and Gofton, "Terminal Dives."
40. Ibid.
41. Proulx, "Review."
42. Ludders et al., "Drowning."
43. Iossa et al., "Mammal Trapping," 344.
44. Iossa et al., "Mammal Trapping."
45. IAFWA, *Trapper Educational Manual,* 98.
46. Inman, "Annual Report," 5.
47. Cole, "Widespread Capture and Mortality."
48. Cole, "Carnage of Wolf Trapping."
49. Proulx and Rodtka, "Steel-Jawed Leghold," 4.
50. Proulx, *Intolerable Cruelty,* 16.

    Proulx also notes that the coyote's mate came after the first struggle bout, stayed with him for short time, then left and never came back. One has to wonder, if she had realized that her mate's fate was sealed and that she could not do anything to help him, or perhaps she knew she was in danger as well.
51. Ibid.
52. Predator Defense, "Trapped Bobcat Rescue."
53. Training Institute on Strangulation Prevention.
54. Funk and Schuppel, "Strangulation Injuries," 42.
55. WHO, "Report," 2.
56. Esteban and Ray, "Measurement," 3.

8

Abandoning Human Entitlement: Empathy, Compassion, and Rights for Nonhuman Animals

Hunter Story

The hunter had just shot a duck out of the air. The bird had crashed onto the ground and he was now towering over her. The duck looked up to him and their eyes met for a brief moment before she tucked her head under her non-injured wing. At this moment, the hunter realized he would have done the same thing—raising his arm over his head to protect himself from a giant towering over him. After this experience, he stopped shooting birds.[1]

Was the hunter struck by a sense of empathy, which is the ability to put ourselves into another's situation to understand their feelings and feel them ourselves? If so, what took him so long? Is his (limited) personal choice sufficient for our relations with other animals?

Emotions

Sportsmen's emotions are captured through questionnaires sent to sportsmen asking about their experience and satisfaction with hunting regulations, which then will be adjusted, usually resulting in expanding hunting

A. Heister, *Beyond the North American Model of Wildlife Conservation*, The Palgrave Macmillan Animal Ethics Series, https://doi.org/10.1007/978-3-031-14149-2_8

opportunities to increase sportsmen's satisfaction. Also, emotional testimonies are frequently offered when sportsmen share their experience of an exciting hunt or along their traplines. Images of sportsmen next to the dead bodies of animals they have killed most clearly convey their emotional states of excitement, exhilaration, and pride. As discussed in the previous two chapters, sportsmen's negative emotions drive policies that lead to the ongoing slaughter of wolves, particularly in Western states. Relations between sportsmen and wild animals involve aggression, violence and killing. In addition, hunters and trappers ferociously defend their assumed right to kill wild animals and frequently express their anger and outright rejection in response to any attempts by the public to curtail or end certain hunting and trapping practices.

Having attended numerous state wildlife meetings over the past two decades, it has been my experience that emotional testimonies by the general public expressing concern for the individual wild animals are seen as naïve and are generally met with entire dismissal by agency personnel and sportsmen alike. Undoubtedly, this is more reflective of sportsmen's inherent speciesism and gendered bias in sportsmen's relations with wild animals than inappropriate, emotional responses by the public. In the world of hunting and trapping, emotions are used to defend the NAM and to describe hunting and trapping as exciting and important human experiences but according to this view, the animals they stalk, assault, injure and kill, do not have emotions and neither are emotions allowed to defend their animal victims.

Thus, when it comes to defining its view of human relations with wild animals, traditional wildlife management insists on the sole use of reason and rationalization. Whereas reason alone generalizes and tries to separate complexities into categories and classifications with rules and regulations, but only with emotions do we grasp the particular of what is right in front of us. In other words, generalizations have their place to help us understand the world around us, but they also omit a closer view of the particular, in this case, of animal suffering and forced death. We see this in the NAM and how it presents us with a course list of cold, detached, and human-centered principles, which effectively hide sportsmen's cruelty and violence against nonhuman animals. Indeed, a lack of emotions can render us blind to the world, making even thinking impossible but most importantly, it prevents us from engaging with the world in a meaningful and responsible way. Specifically, not only are emotions at the root of and the driving force behind our thoughts and actions but they, not reason alone, make us care about what happens in the world. As Finnish animal ethicist, Elisa Altola writes, emotions are the source of moral concern, "to which reason only contributes fine-tuning: we care because of emotion, and arguments only sculpt what we already know."[2]

In its most dangerous form, reason alone blocks emotions such as empathy and compassion. Emotions are essential for both, our sense of meaning and in our sense of morality and suffering. Here, the link between emotions and recognizing animal suffering emerges, whereby we not only see an animal before us, but we actually feel for her, with her, taking in her experience. Whereas intellect alone can block emotions and lead to disengagement, emotions such as love, tenderness, empathy, and compassion fill in the moral details of a particular situation of someone in need, for example, a deer about to be shot by a hunter. Furthermore, these pro-social emotions are the core elements of understanding suffering and the animal's experience of what is about to happen to her. If we instead turn to theories and regulations, and kill the individual 'for the greater good,' (the species or to improve human life), without emotions, we no longer recognize and perceive the moral relevance of the animal's experience of pain, suffering, and forced death. In general, it is not theory and abstractions but emotional reactions ranging from disgust, anger and outrage in response to injustice and unfairness that motivate us to become advocates and speak up for others. This is what motivated me to act and help the trapped pine marten mentioned in the Prologue of this book. I felt such a strong urge to right this injustice against wild animals that it spurred me into the action of co-founding Footloose Montana to fight for an end of trapping on public lands. It makes sense that a lack of emotions causes us to turn into monsters, while the cultivation of pro-social emotions has the potential to span a solidly kind web and connect detached people, and perhaps even helps us with bridging the artificial gap between us and other animals. In her explanation of the importance of evoking emotions in animal ethics, Aaltola writes that theory alone can keep people cold but when accompanied by emotions, this can help us to grasp the obviousness of what is wrong with causing other animals suffering.

Research in psychology and neuroscience supports that deliberate decision-making involves and is improved by both cognitive processes as well as quick emotional responses. We actually expect other humans to react with anger and outrage when confronted with injustice or unfairness, and often we perceive others who display a lack of emotions, in particular, an absence of empathy as callous and 'cold.' In her book *The Sociopath Next Door*, psychologist Martha Stout (2005) writes that a lack of empathy is symptomatic of an absence of conscience, which in turn characterizes an antisocial personality disorder that exists in approximately 4% of the U.S. population (this fraction of the population is representative of conscienceless sociopaths, who have no empathy or affectionate feelings for humans or other animals). This also means that people who lack empathy are most likely to treat others as objects. It also indicates the profound need for empathy, compassion, and humane education training. In sum, research findings confirm that emotional reactions to injustice are normal and that emotions are essential for making appropriate judgments and decisions.[3]

Already in the 1800s, Charles Darwin recognized the crucial role of emotions as essential for the survival of human and nonhuman animals. Emotional display plays a crucial role in safety and survival. For example, we read the body language and behavior of others to assess whether we are safe or not. Thus, emotions help us to take a shortcut and circumvent thinking in dangerous situations where we need to act quickly and decisively, which increases our chances of survival. In fact, emotions (affection, joy, sadness, anger, anxiety, aggression, fear, and so on) suggest a form of language and communication[4] that is independent from the ability or inability of the spoken language, which allows us to read the emotional state of humans and nonhuman animals.

Beyond the North American Model of Wildlife Conservation, the value of emotions is discussed and viewed as a source of moral understanding, in particular as part of the approach of compassionate conservation. This new thinking aligns with the current scientific and philosophic understanding that reason and emotion complement each other and are better understood not as opposites but as partners.[5]

Empathy

Empathy is of particular importance in our relations with others, humans and nonhuman nature alike. The term derives from the German word *Einfühlungsvermögen,* which means the ability to *feel into.* The cluster of empathy entails sympathy and compassion (German *Mitleid—feeling with*) and they are often used interchangeably. However, there are important differences whereby the standard definition views *sympathy* as feeling *for* someone, while *empathy* is feeling *with* someone. Expanding on this, empathy is an approach through which one can recognize, understand and share the experience of another human, a nonhuman or even fictional character.

Today, empathy is understood as an umbrella term in psychology covering three distinct, yet related abilities—affective empathy, cognitive empathy, and empathetic concern, which occur in different parts of the brain but all help us toward understanding the perspective of another. The first, affective empathy enables us to sense or physically experience the perceived emotions of others—a hardwired process that involves mirror neurons. This is how we automatically respond sympathetically to crying or contagious yawning. An affective empathy response may be related to biophilia, our innate tendency, and affinity toward the natural world, which is important as a powerful building block for establishing care for nonhuman world.[6] Cognitive empathy allows us to mentally understand what someone else is feeling and why. Finally, empathetic concern that can also be called compassion has an inherent motivation for action to relieve the other of her/his suffering. Here, we feel motivated to help and relieve the suffering of a nonhuman animal in distress or a struggling human.

We can learn to recognize and value the understanding that moral worth exists in both the individual *and* the whole they belong to. While the reliance on reason alone facilitates competition and hierarchical thinking, if we instead allow positive feelings such as empathy and compassion in our interactions with nature, we can overcome these dualisms and dissolve these opposite positions. Taking into account that the border between individual organisms and ecological communities is blurry, and that respect is the foundation of concern for each, empathy can serve to transcend both. Accordingly, empathy plays a crucial role in merging the concern for the individual animal (usually the focus of animal rights advocates) and concerns for animal populations (generally expressed by conservationists).

The power of empathy can help us to understand the experience of another and feel with the other. With respect to wild animals, we may cultivate empathy and compassion to come to a better awareness of what other beings are going through. For example, we may realize that animals' lives are often already burdened with difficulties and challenges and that we should not add yet another layer of risk of mortality for them through hunting and trapping. We could then feel the grief and sadness in the surviving partner, missing her mate who was killed by a sportsman. We could imagine the despair and particulars of the slow and painful death of dependent young left behind in the den who starve to death. This empathetic approach could lead us to compassionate action for individual animals set in motion not only by concerns to prevent pain and suffering but beyond that, wanting to make sure that wild animals are flourishing and supported in their lives. The importance of empathy cannot be overstated as part of humans' ontology and with respect to our relations to nonhuman nature. Not only do empathy and compassion help us in developing our own moral awareness and thus contribute to individual transformation, but they are also necessary requirements for broader social change. Here, pro-social emotions such as sympathy, empathy, and compassion are thought of as contributing to behaviors intended to benefit another, such as helping, sharing, and comforting. They are considered the social glue of society and are crucial for our relationship within our own species and with the more-than-human world.

Shared Experiences: Empathy as an Intra- and Interspecies Connector

The literature is filled with examples of behavioral observations of wild animal species that reflect shared experiences between humans and nonhumans. These include observations that range from elephants grieving over the loss of their own, New Caledonian crows making sophisticated tools, to wolves, coyotes, and foxes, with species-specific moral codes that guide their intra-species behavior—such as the concepts of fairness and altruism.

Empathy can be viewed as a connector among individuals within populations, as well as across the nonhuman and human species line[7] with significant implications for science and ethics; first, empathy is an example of a shared skill that has developed as part of co-evolution, and empathy is tied to sentience, both of which the best available science must include; second, empathy and compassion are important human characteristics that need fostering within human society.

Renowned primatologist Frans de Waal and colleague Stephanie Preston have made us aware of 'a shared-state mechanism' through which we perceive and understand another's emotional state, essentially an intersubjective experience termed "emotional linkage."[8] Referring to these researchers' work on "The communication of emotions and the possibility of empathy in animals," Marc Bekoff and Jessica Pierce further explain that empathy is not a single behavior but, instead, describes a whole class of behavior patterns that exists across species and shows up with varying degrees of complexity. As evolution builds complex from simpler forms and (re-)uses existing structures and processes, emotional linkage, especially between mother and infant, likely evolved into emotional contagion and from there, to more complex forms such as cognitive empathy. Nested levels mirror existing brain structures, with the inner core a necessary foundation for the other layers. The inner core consists of relatively simple forms of empathy such as body mimicry and emotional contagion (e.g., yawning, or the joint crying of human babies in nursery following the first, or birds flying off as soon as the first startled bird does), which are largely automatic physiological responses. The next layer consists of somewhat more complex behaviors such as emotional empathy and targeted helping. More complex still is cognitive empathy, or the ability to feel another individual's emotion and understand the reasons for it. Finally, and most complex is the capacity for attribution, in which an individual can fully adopt the other's perspective, using the imagination.[9] Since empathy is part of the same neurological architecture as other pro-social behaviors such as trust, reciprocity, cooperation, and fairness, it seems likely that a whole suite of interlinked behaviors have co-evolved in social mammals, again, confirming Charles Darwin's statement that moral behavior in humans and nonhumans comes in degrees and not kind.

Research conducted by veterinarian and ethologist Michael Fox has shown that wolves display many more facial expressions than coyotes and red foxes. As such, Bekoff and Pierce surmise that wolves also potentially have developed more complex "moral capacities and nuanced empathy, perhaps displayed as communicating and responding to more subtle variations of social rules, than coyotes or red foxes." The authors have further argued that facial expressions (and the ability to read them) are linked with social and moral complexity, explaining that "the more facial displays, the greater and more nuanced the social information that can be communicated."[10] After all, we come to a perception and understanding of others' emotional state via reading people's expressions (facial and body language).

Anecdotes and empirical evidence suggest that empathy at least exists in elephants, several cetacean species (especially bottlenose dolphins and toothed whales), rats and mice, social carnivores, and primates."[11] And then there is always the unknown, for example, what prompted a group of crows who were observed circling around other dead crows in a funeral-like fashion, or, were they perhaps trying to investigate the cause of death of their comrades?[12]

Implicit in these research findings are implications for the treatment of other animals in science and other domains of human society. Not only are other animals sentient beings like humans, but if other animals share with humans the capacity for empathy, then they also "have in place the cornerstone of what we in human society know as morality."[13] Indeed, behavioral biology and neuroscience have proven that cooperation is very common in other animal species, whereby animals engage in moral behaviors and react to inequity, which is linked to morality (this has been discussed in Chap. 2). From food sharing to reconciliations, conciliation, and conflict mediations, these methods of resource distributions and conflict resolutions, known to be the very building blocks of morality often involve "the capacities for empathy, sympathy, and even community concern."[14]

To conclude, if empathy is spread across species lines and serves in humans as a prerequisite to be compassionate, reduce pain and suffering and do good for others, these and other shared experiences need to be considered in science and ethics as components of the animal standpoint.

Barriers to Empathy

One mechanism of how anthropocentrism blocks our empathy is its binary logic associated with a hierarchy of moral worth. One could easily assume that human supremacy over nonhumans directly leads to concern for all humans, but the evidence suggests otherwise. In fact, research by Canadian political theorists and animal ethicists Will Kymlicka and Sue Donaldson shows that people's attitude of differentiating between humans and animals is positively associated with dehumanizing human outgroups such as immigrants. In turn, people who recognize valued traits and emotions in animals are also more likely to accord equality to human outgroups. From this, Kymlicka and Donaldson concluded that reducing the status divide between humans and animals helps to reduce prejudice and to strengthen belief in equality among human groups.[15]

A growing number of psychologists and scientists presume that humans will naturally develop empathy unless thwarted by external forces. But if empathy is a natural response to suffering, why then does it so often fail in our relations to other-than-human animals? Many theorists have noted that ideological conditioning plays a major role in blocking the natural pathways to compassion, the common notion of 'it's just an animal' being just one of them. As previously discussed, the biblical mandate that allegedly gives humans 'dominion over all the earth' has been very detrimental to the relationship between humans and other nonhumans (wild and domestic). These social and cultural beliefs are powerful forces and implicit in our individual and collective strategies to avoid feeling empathetic with other animals.

Feminist philosopher and animal rights advocate Brian Luke has examined four personal and collective strategies which people use to block their empathy for nonhuman animals: (1) "denying personal responsibility," (2) "denying the harm done," (3) "denying animal subjectivity," and (4) "overriding sympathies for exploited animals."[16] The pervasive existence and power of these blockades point to the critical value of cultivating pro-social emotions to overcome them. This also emerges when we consider the intrinsic value of others, regardless of whether that is another human being, nonhuman animal, or nature. Providing an example from our human relationships, philosopher and environmental ethicist Katie McShane explains that, if we say that we love a friend because she serves our interest of not being alone, this means that we deny her value apart from our interests. Here, McShane notes that love involved in relationships such as friendships and our human relations with the more-than-human world is an '*other*-centered' emotion.[17] In the same vein, the value of other animals does not lie in the question of what they can do for us, but in their being apart from our interests. Truly feeling connected to, respecting, and loving other beings means to care about them despite, or perhaps *because* of their independence and often indifference to us; a beaver may have instrumental value because of his water engineering capabilities that ensure water availability for ecosystems, other animals and humans (ecosystem services), but he has also intrinsic value as a being, and he matters to his mate and family. With this recognition, it becomes rather unconscionable to treat nonhuman animals with total disregard for their wellbeing. Furthermore, acknowledging the intrinsic value of nonhuman nature de-centers humanity from the moral universe and embeds humans within a complex biosphere of others with whom we engage in moral relationships.[18] Anthropocentrism is incompatible with human attitudes of love, connection, and respect for nonhuman nature.

In part through the pervasive and self-serving propaganda by corporate industries that exploit nonhuman animals, such as the livestock industry, 'meat'-producing corporations, and the Conservation-Industrial Complex promoting hunting and trapping, we hold on to the culturally ingrained belief that what defines a good life for humans requires the death of nonhuman animals. In feminist Josephine Donovan's words, "To a great extent … getting people to see evil and to care about suffering is a matter of clearing away ideological rationalizations that legitimize animal exploitation and cruelty."[19] To this end, becoming aware of our own attitudes, use of language, along with the choices we make, is a positive step toward granting nonhuman animals the same social considerations we innately give other humans. Importantly, there *are* different forms of ways for us to be in the world. We can shift our awareness toward a different, inclusive ontology and embrace the fact that both humans and nonhumans are denizens—"conscious, feeling, communicative selves, bound to other conscious selves through various webs of relationship and dependency, each with our own subjective experience of the world. The human good is now continuous with that of other animals."[20]

Choice of Language

Previous chapters have already addressed some of the harmful and empathy-suppressing terms of categorizing wild animals, from 'game,' furbearers to 'natural renewable resources.' In addition, Colorado-based education coordinator Ashley Young and colleagues explain that barriers of language choice and narrative framing surrounding animals and their experiences can be powerful tools to either promote or dissuade empathy. For example, referring to a nonhuman animal as "her/him/them" instead of "it," helps us understand the nonhuman animal as an individual with a particular perspective (rather than just an object). By categorizing animals as 'pets' to be cared for, while others are 'pests,' and so on, sends conflicting messages throughout society, especially to children.[21] Another example is the dissonance caused by wildlife agencies when they, on one hand, encourage the public to 'save bears' by picking up and securing their trash and fruit around their homes at the end of the summer, while on the other hand, promote the hunting of bears.

Too Much Empathy?

Nevertheless, Young and colleagues caution that there is the potential of eliciting too much empathy, which can lead to disengagement. They explain that "when people are overloaded with highly emotionally triggering experiences, they may express moral disengagement by justifying their negative actions, shifting the blame to remove personal responsibility, disregarding the consequences as out of their control, or dehumanizing the victims."[22] This amplifies the importance for us to learn how to set boundaries. Becoming aware of boundaries is part of the field of emotional intelligence—our capacity to be aware of, control, and express one's emotions, and to handle interpersonal relationships judiciously and empathetically. Training in EI can help us to become more familiar with our inner emotional landscape, triggers, and how to respond to life's challenges, including the facing of injustices against nonhuman animals and how to respond to the interlinked ongoing crises on a global level.

The Need for a Non-anthropocentric Ethical Foundation: A Global Perspective

It is a grim reality that human corporations and private special-interest groups are globally destroying wild animals and wild places at a catastrophic level, and two recent findings speak volumes; The 2020 Living Index commissioned by the World Wildlife Fund for Nature (WWF) documented a 68% decline in the sizes of birds, mammals, fish and reptiles since 1970.[23] This report was preceded by yet another stark warning by the 2019 United Nations-backed landmark report from the Intergovernmental Science-Policy Platform on Biodiversity and Ecosystem Services (IPBES) on the rapid decline of nature, warning that the interconnected web of life is getting smaller and increasingly frayed and that one million animal and plant species are in imminent danger of extinction, many within the next few decades.[24]

Both reports identified human activities as causes and highlighted the need for transformative social changes and values. Neither report, however, addressed the anthropocentric paradigm that humanity is operating under, a most destructive belief system that assigns an exclusive role and center-place to humans, while viewing wild animals and their habitats as inferior, and pushing them both to the periphery of moral concern.

In addition to blatant extinctions, the killing of wild animals is so profound that we are literally emptying out the planet, a human-caused phenomenon called 'defaunation.' Sociologist and author, Eileen Crist has further warned us that our human supremacy thinking is even prioritizing climate change over massive species extinctions. Crist notes:

> [But] the reason climate change has penetrated public awareness is because an overheated planet directly threatens humanity and civilization. Mass extinction, the unraveling of the web of life, isn't seen as so grave, because it isn't viewed as an existential threat to us. It's happening to *them*: the insects, the fish, the frogs, the birds. Putting aside the fact that these animals have inherent worth, we are making the typical mistake of thinking that humans are somehow separate from, and not dependent upon, the Earth's natural systems.[25]

Humans make up less than one percent of approximately 8.7 million species (6.5 million species on land and 2.2 million in oceans), yet, at a global population of 7.8 billion with a steep trajectory to be at 9 billion by 2050, humans now already use more than 25 percent of all the photosynthesis on Earth. "Changing land use for food production is the biggest driver of nature loss. About 50% of the world's habitable land area is already used for agriculture—for livestock such as cattle and pigs, and for crops that feed both people and livestock."[26] Not only has 'meat' consumption been identified as a leading cause of the loss of biodiversity but also as a major contributor to climate change. The pervasive belief in 'meat' as the only or the best protein source is highly detrimental. The reliance on wild animals for food, in particular the 'hunting for food' mentality, has been shown to be the leading cause of the loss of megafauna globally. As the results of a 2018 study suggest, we are literally eating the world's megafauna to extinction, and not only that but human demand for wildlife as food also threatens the survival of many other animal species, including birds and fishes.[27]

In contrast, a study found that a vegan or vegetarian diet had the smallest carbon footprint compared to the production of animals for food and recommended the reduction of meat consumption, including so-called 'wild meat' as much and as soon as possible.[28] Indeed, switching to a predominantly plant-based diet is one effective solution to help shrink land, freshwater, and carbon "hoofprint" to combat mass extinctions.[29]

It has become worrisomely apparent that all species on planet earth have arrived at a pivotal point in our shared evolution. As wild animals and their homelands are edging closer to extinction, the urgency to change our relationship with nonhuman animals is rapidly increasing, bringing the need for a non-anthropocentric ethical guide for conservation (and for our lives) squarely to the foreground. The next section provides an introduction to Compassionate Conservation, an approach to conservation that, to my mind, provides an effective path forward in our efforts to protect biodiversity while reducing human-caused harm and suffering of nonhuman nature. The focus on compassion as a moral virtue that serves as a guide for conservation is a reminder of our human potential to do good in the world, and with that, it has the potential of replacing the contemporary lethal paradigm of the North American Model into a humane and just one.

The Urgent Need for Compassionate Conservation

Marc Bekoff, among others, has pointed to compassion as a must for advancing global conservation. Specifically, as he and colleague Daniel Ramp assert, "A compassionate conservation approach aims to safeguard Earth's biological diversity while retaining a commitment to treating individuals with respect and concern for their wellbeing."[30] We may think of a compassionate conservationist as an individual with the disposition of good character characterized by the ability to be receptive and responsive. Also, given that compassion is a moral core virtue of the world's major philosophical and religious traditions, we can conclude that it is appropriate for conservationists (along with anyone else) to demonstrate compassion with regard for the lives and well-being of ecosystems and animals. "A virtuous person will carefully attend to the capacity of others to experience both joy and pain and make efforts not to inflict intentional and unwarranted suffering as a manifestation of one's compassionate character."[31]

Concerns by the public, conservationists, and animal rights advocates alike have led to calls for a new paradigm in conservation. Initiated by Will Travers, president of the Born Free Foundation, a subsequent conference at the Oxford University in the UK, brought together the two disciplines of animal welfare and conservation that are often perceived as mutually exclusive, and articulated the new approach of Compassionate Conservation. Since its inception in 2010, Compassionate Conservation has gained momentum, sparked the interest of some non-governmental organizations, and established the Centre for Compassionate Conservation at the University of Technology, in Sydney, Australia. This approach brings empathy and compassion into the decision-making and policy-development process by emphasizing that individual wild animals are not objects or commodities who can be traded off for the alleged 'good' of their own or of other species' nor for the 'good of ecosystems.[32] A leading view by compassionate conservationists is that conservation should expand its moral community by recognizing all sentient beings as persons. As further explained by Wallach,

> Personhood in an ethical sense implies the individual is owed respect and should not be treated merely as a means to other ends. On scientific and ethical grounds, there are good reasons to extend personhood to sentient animals, particularly in conservation. The moral exclusion or subordination of members of other species legitimizes the ongoing manipulation and exploitation of the living worlds, the very reason conservation was needed in the first place. Embracing compassion can help dismantle human exceptionalism, recognize nonhuman personhood, and navigate a more expansive moral space.[33]

With growing awareness of other animals' sentience and sapience, "standard conservation practices that categorically prioritize collectives without due consideration for the wellbeing of individuals are ethically untenable."[34] Given that the majority of problems addressed by conservation are human caused, why do wild individuals having to pay the price with their lives for our insufficiencies and shortcomings? Too often trade-offs are part of conservation projects, where decisions are being made which animals get to live and who must die. For example, trade-offs often entail killing members of one species for the benefit of another (e.g., golden hamsters for black-footed ferrets).[35] Other trade-offs involve research projects whereby some members of one species are killed to allegedly benefit another. Such was the case in a 2014 Canadian research study[36] during which 890 wolves were mass killed; wolves were aerially gunned down, trapped, asphyxiated in neck snares, poisoned with strychnine while others were outfitted with judas collars. The 'collateral damage' involved the poisoning of 91 ravens, 36 coyotes, 31 foxes, 8 marten, 6 lynx, 4 weasels, and 4 fishers. Not only did the mass killing of wolves not lead to the desired objective of reducing the decline of Canadian Boral Caribou, for which wolves were blamed, but the study exposes an appalling callousness and disregard displayed by the researchers, for the lives and stress and horrible suffering and deaths caused by first chasing wolves with planes and using traps, snares, and strychnine (known to cause extreme suffering). The issue of the precarious decline of caribou is much more complicated but it starts and ends with humans, and related economic activities such as hydroelectric installations disturbing the historic balance between wolves and caribou and causing an average loss of habitat across caribou ranges more than 70 percent.[37] Other 'trade-offs' accept that some individuals will die as a result of translocation projects, as was done in a 2020 Montana project where 30 pine marten were caught in live traps and moved from their home to a different, unfamiliar area with the expectation that the *species* would get established. Projects like this are conducted without measures taken to ensure that all pine martens will survive or at least be tracked to stay informed about their welfare. In contrast, such a translocation project under the framework of Compassionate Conservation asks us not only to take into account the species-specific needs as well as to consider the best season, where the best habitat is to release these animals into, and the best methodologies and mechanisms guiding their release. In sum, Compassionate Conservation demands that we know *who* we are translocating, which requests a range of requirements, including careful planning, and assessments of stress levels, risks, and individual personalities of the animals to ensure the welfare of individuals involved, which overall should lead to better conservation outcomes.[38]

The Four Tenets of Compassionate Conservation

From its overarching key principle—treat all beings with compassion and mitigate human-caused suffering—flow the four tenets that have been promoted by Will Travers, Marc Bekoff, Dror Ben-Ami, Dan Ramp, Arian Wallach, Michael Nelson, Chelsea Batavia, and many others who are deeply committed to advancing Compassionate Conservation. The principles are, "First Do No Harm," "Individuals Matter," "Inclusivity of all wild animals," and "Peaceful Coexistence." Each of these principles is a specification meant to guide conservation practitioners in their actions based on the primary value of compassion.[39] *First, do no harm*, is adopted from the traditional tenet of medicine, the Hippocratic Oath, and means that our tendency for human interventions should be carefully reflected upon and scrutinized. This includes the option of not doing anything at all when it is determined that intervening would risk more harm than good. The principle stating *individuals matter* acknowledges the intrinsic value of individual wild animals and rejects killing them or reducing their value to mere parts of the collectives; *Inclusivity* acknowledges the intrinsic value of both all wildlife individuals and the collectives they belong to, and regardless of the size of their populations, whether their ancestors were introduced, or native, whether they are considered sentient or not, and regardless of their usefulness to humans. Finally, *peaceful coexistence* calls for recognition that the first instinct in conflict situations should be to critically examine and modify one's own practices, rather than pursuing acts of aggression [and violence] against individual wild animals.[40] These tenets serve as an aspirational guide for our interactions with individual wild animals while we simultaneously strive toward the protection of collectives. "In practice, a compassionate conservationist works to develop, apply, and prioritize nonlethal and noninvasive strategies that benefit wildlife collectives without causing intentional suffering to wildlife individuals."[41]

Critique of Compassionate Conservation

Not everyone agrees with compassionate conservation. In fact, its introduction into the conservation community has been met with strong and negative responses, some of which seemingly fear that allowing animal subjectivity to influence decision-making would taint their science. Here the argument provided often is that compassion, sentimentality and anthropomorphism have no place in conservation. However, other conservation scientists object to this exclusion, for example renowned field biologist George Schaller stated in an interview that not allowing emotion makes for a dead study. As field biologists are "dealing with individual beings who have their own feelings, desires and fears," he recommends using some emotional contact and intuition.[42]

Other concerns about Compassionate Conservation have been expressed, some of which center around two main issues: the paradigm change from a utility-based practice to a morality-based practice, and second, objections against viewing non-native, 'invasive' animal species as natural. These are not easy topics, but this is one of the arguments made in support of Compassionate Conservation, namely that it asks (and responds to) difficult questions and addresses topics to move us out of our comfort zone. For example, Compassionate Conservationists point out that a de-categorization of wild animals is essential to improve the lives of wild animals. Categorization of animals into "overabundant" or "nuisance" or "invasive" shape and perpetuate our often-negative attitude and relationships with them. There is a noticeable lack of consideration of the moral relevance of the sentience of members of species termed invasive when deciding on the best possible management strategies.

There is also a clear value bias against invasive species, where such species are viewed and presented as inherently "bad." It is likely that because of this almost exclusively negative view, many wildlife managers, conservation biologists and others involved in the control of invasive species management, share a unified bias that these animals are "pests" and must be removed or eradicated by the most cost-effective and efficient methods available. These are difficult topics. For example, there is no dispute about the ecological devastation that cane toads cause, a species introduced to

Australia, where they are contributing to the further decline of species already in peril such as the 'native' quoll populations (marsupials), who mistake these poisonous toads for food. Another well-known example involves Burmese pythons in the Everglades in Florida, who present an unknown predator to many of the resident and 'native' animal species, including mammals, birds, and reptiles. Consequently, many species such as raccoons, opossums, rabbits, bobcats have declined significantly. Originally, the snakes were likely victims of the pet trade and 'released' into the Everglades. Since then, they have been rounded up for years, persecuted, and killed by sportsmen and agency personnel as part of organized killing events. The neglect or omission of sentience within traditional conservation often carries over into the implementation of invasive species management techniques, where sentient members of invasive species are often not treated humanely. This is tragic, and given that the majority of these species are the result of human introductions, why should these animals pay the price for our carelessness? However, instead of immediately resorting to calls for 'eradication' of the introduced animals, these situations challenge us to ask questions about human accountability, our responsibilities with respect to human-caused harms, and to find more ethical and modern solutions.

We can no longer eliminate our own species and our exploding populations with all the problems linked to it, from the discussions on these types of human-caused 'disasters,' or the detrimental effects of our own 'invasiveness.' Considering our own destructive forces in the process of discussing (ecological) problems may help us find more creative solutions grounded in compassion and with a sense of justice and fairness toward nonhuman animals. In the meantime, steps that should be taken immediately, are enacting laws that prohibit the importation and exportation of wild animals, any 'ownership' of wild animals, including wild animals as pets. These divisive actions require us to change our 'human supremacy' belief system and codify the personhood of nonhuman animals into the law, among many other structural changes. As a sufficient treatment of this debate falls outside the scope of this book, and many of my own arguments in support of Compassionate Conservation would repeat many of the topics addressed throughout, I refer the reader to initial publications on Compassionate Conservation such as *Ignoring Nature No*

More: The Case for Compassionate Conservation (2013) edited by Marc Bekoff, and the subsequent discussion that entails the critical perspective against CC by some conservationists and multiple responses by CC conservationists addressing critics' concerns in a decisive and convincing way. "Compassion as a Practical and Evolved Ethic for Conservation" (Ramp and Bekoff 2015), "Summoning compassion to address the challenges of conservation" (Wallach et al. 2018), and "Recognizing animal personhood in compassionate conservation" (Wallach et al. 2020) were especially helpful in the research process that led to the writing of this book.[43]

The Need to Replace the NAM with Compassionate Conservation

By challenging the socially constructed, superior status of humans, Compassionate Conservation provides an impetus for a different belief system and thus a vision of better relations between human and nonhuman nature. With respect to the NAM, it does so by addressing three of the most detrimental aspects of traditional wildlife management:

First, it eliminates our acceptance of wild animals as 'natural renewable resources,' and instead sees and treats them as other persons with interests and needs that need to be fully protected. Their elevated status would change our relations toward a more respectful and ethical perception that also considers the interdependence in an interconnected world. Consequently, management strategies would move in a different direction with different goals and objectives. It would recognize that the diversity and health of wild animals require more careful, thoughtful, and empathetic approaches.

Second, a recognition of the independence and autonomy of wild animals could modify the ongoing transition of the public's changing relations toward mutualism with wildlife. For example, Canadian philosophers Sue Donaldson and Will Kymlicka suggest that wild animals, just like humans, are citizens and capable of exercising their own sovereignty in negotiating their environments.[44] This thinking and the view of wild animals as independent and autonomous agents surfaces in other current concepts such as property rights, and the movement to codify individual animals as legal persons. An equal platform such as this would most certainly not only protect individual animals much more effectively from harm and violence but would also provide a more ethical springboard from which to develop more equal and just relations within the context of kinship.

Thirdly, as suggested by Max Foran, we could rely on the very qualities that we claim to differentiate humans from other animals. "We can resort to our higher levels of consciousness and ask ourselves what our rational responsibilities are toward those whom we are able to dominate?" And as philosopher Paul Taylor has succinctly noted: "When rational, autonomous agents subscribe to the principles of moral consideration and intrinsic value and so conceive of wild living things as having that kind of worth, such agents are *adopting a certain ultimate moral attitude toward the natural world* [that] … parallels the attitude of respect for persons in human ethics."[45] Is it then not a rational perspective—or a way of being—to recognize and respect the autonomy of wild animals on a planet to which nobody has exclusive rights to? And, if we consider our most important traits, we can empathize with the challenges of wild animals. When we are compassionate, we are motivated to help others in reducing suffering as well as forming and maintaining cooperative relationships. Conversely, killing animals dismisses their intrinsic value and violates this bond and our determination to mitigate human-caused harm.

Positioning compassionate conservation alongside traditional wildlife management decisions would make the costs to wild animals visible and thus rejectable. Compassionate conservation de-centers the human experience by viewing nonhuman animals on par with humans, whereas the current approach prioritizes human interests and exhibits callousness toward the individual animal. Compassionate conservation also expands the moral community while the NAM insists on a myopic focus privileging sportsmen and remains in service to special interests. And finally, human relations with nonhuman nature as envisioned by Compassionate Conservation is motivated by pro-social emotions, whereas the NAM entails aggression, violence, and killing.

An example of how Compassionate Conservation can change lethal management practices involves a small group of Little Penguins on Australia island whose survival was threatened by foxes. Lethal measures were used to kill the foxes through poison and trapping. In 2006, a program was implemented where Maremma sheepdogs now guard the penguins, and human members of the local city council visit the dogs daily and walk with them around the island to spread the dogs' scent and deter the foxes. Compassionate conservation allows for—but does not prescriptively dictate—outcomes in which the interests of others supersede those of humans. As such, this approach can be seen not only as a moral imperative for providing modern solutions for sharing space with nature but also for fostering the possibility for diverse species to live in peaceful coexistence.

Project Coyote, a California-based nonprofit organization provides an example from North America demonstrating that human communities can peacefully co-exist with predators such as coyotes who are increasingly entering urban settings. Where human-coyote conflicts usually resulted in coyotes getting killed through trapping and poison, methods that are not only unethical but also ineffective, Project Coyote has been able to redirect these lethal methods toward public education where people learn how to reduce the risk of harmful encounters with urban coyotes.[46]

Rights for Nonhuman Animals: The Nonhuman Person

Only the legal status of personhood will afford non-human animals rights and protection. Ongoing efforts by Steve Wise, animal rights attorney, and director of the Nonhuman Rights Project in the US, are aimed at obtaining legal personhood for certain nonhuman animals. Wise argues that gaining an individual's or a species' legal standing is related to abilities associated with sentience (consciousness etc.) Wise has been leading the legal fight to free several incarcerated individual animals, including

Tommy, a chimpanzee, who has been kept in a cage in a room in a warehouse in Johnstown, New York, for many years.[47]

Notions of animal welfare inherent in some regulations are insufficient in stemming human violence and preventing animal exploitation. An expansion of rights to include rights for wild animals is not very far-fetched and many efforts have already achieved or are in the process of achieving rights for nonhuman nature. Inspired by the late theologian Thomas Berry who proposed that "the Earth is a communion of subjects, and that rights originate where the universe originates and not from human jurisprudence," many efforts are underway to overcome our ingrained sense of human supremacy. As explained by South-African environmental attorney and author, Cormac Cullinan, this means "we cannot claim that humans have human rights without conceding that other members of the Earth Community also have rights."[48] For this to happen, nature and its nonhuman denizens need to be released from their legally enshrined property status. Instead, the more-than-human world must be recognized as having rights to exist, persist and flourish, with people having a moral obligation and authority to enforce nature's rights on behalf of ecologies and their denizens.

Writing and discussions about this enormous transformation of our relationship with nature has also been taken up by many scholars and citizens as part of the rights and liberationist movement as well as the Rights of Nature movement and associated legal initiatives such as the Earth Law Center. Beginning in a borough in Pennsylvania in 2006, there are now seven countries, including Australia, Colombia, and Ecuador, which have included rights of nature in their constitutions. Most recently, the Maori tribe in New Zealand have achieved the legal recognition of a large river as an ancestor with legal personhood.[49] This movement provides us with a much-needed holistic ethical and legal framework that re-embeds humans into the ecological context and gives nature a voice. This justice-based vision of Earth democracy has already begun to shape a crucial egalitarian relationship with the more-than-human world.

Property Rights for Wild Animals

In recognition of the autonomous nature of wild animals, we can also develop an awareness of them as citizens, or as property owners. As previously mentioned, Will Kymlicka and Sue Donaldson have put forward the view that wild animals, just like humans, are citizens and capable of exercising their own sovereignty in negotiating their environments. Their thinking entails a hands-off approach and recognizes the skills of wild animals in managing their own lives in the face of daily risks and challenges they experience. Here, wild animals are viewed as "competent actors in a division of labour in which they take responsibility for mutual assistance in their own communities, and indeed are much more competent to do so than we would be on their behalf."[50] An inspiring hands-off approach has already been practiced by the Canada-based Raincoast Conservation Foundation, which, for example, purchases land and trophy hunting tenures, which gives them the exclusive right to guide trophy hunters, which they will not do. Instead, guided by science and compassion, the foundation protects coastal wild areas with their wild animal inhabitants, prohibiting trophy hunting and trapping and putting the interests and the wellbeing of wild animals first.[51]

Property rights for wild animals has been promoted by Australian philosopher John Hadley, who argues for specific and suitable ecological areas being set aside and given title as the property of the animals to be held in trust by a guardian. Harm in any way would be a legal violation of property rights. In this way, Hadley argues, "The nonhuman property regime … strikes a balance between the naturalism explicit in environmentalism and a naturalistic orientation implicit in animal rights liberalism."[52] This idea has been extensively explored by Karen Bradshaw, in her 2020 book "Wildlife as Property Owners: A New Conception of Animal Rights." Her main argument for a radical, interspecies system of property is compelling and urgently needed. Bradshaw proposes that the law allows wild animals to own land through a trust, the same way many humans already do. The trust would be managed by a human trustee for the benefit of wild animals, the beneficiaries. Under her proposal, Bradshaw explains, human trustees would manage land for animals at an ecosystem level, with a mix of public and private judges overseeing their work to ensure that they act in the best interest of animal beneficiaries.[53]

Ecodemocracy: Nonhuman Animals Need Representation

More efforts are underway to help abolish our human-centric worldview and create a better society for all animals—humans and nonhumans. Scholars like British animal ethicists and writers, Joe Gray and Patrick Curry have started to create a political process that recognizes the intrinsic value of non-human nature, and how to put 'ecodemocracy' into action. The thinking here is that, according to established conceptions of democracy, all interests call for equal consideration in the democratic decision-making process. The challenge, however, would be how to decide which humans could serve as proxies for nonhuman nature. Certain qualifications are needed to make effective moral judgments on behalf of other life forms. Through science and experience, humans could increase our understanding of other species and maybe even uncover what they want. After all, it is not that other animals do not express their preferences and votes. With awareness of an element of interpretation and the difficulty inherent in speaking for another being, feminist legal scholar and author, Catherine MacKinnon writes that animals do "dissent from human hegemony," noting that "they vote with their feet by running away. They bite back, scream in alarm, withhold affection, approach warily, fly and swim off."[54]

Other thinkers such as the late Australian philosopher and feminist, Val Plumwood (previously introduced) have urged that social and ecological justice must proceed hand-in-hand and that support for implementing human speakers representing nonhuman's interests would ideally come from leaders such as sympathetic politicians, activists, educators, public intellectuals, and workers in all media[55] (and I would add, conservationists and attorneys). Imagine if conservationists and other (empathetic) members of the public, properly trained in ecodemocracy, would be added as members of the boards of national parks, city councils, and other public executive committees and be the (voting) voices of nonhuman animals and other parts of nature. Imagine how such an inclusive political decision-making process could (it is hoped) reduce or even prevent much of the harm to non-human entities.

Acknowledging the difficulties of implementing a democratic process into an anthropocentric system, Gray and Curry suggest a kind of deliberative democracy where a number of humans would be specifically tasked with speaking for non-human nature in the discussions, to ensure that the intrinsic value of non-human nature is considered as well as to balance out any anthropocentric ideas. Both scholars emphasize the need to prepare the human representatives, a practice that should include background reading and solitary reflection or meditation, combined with group-based expressive exercises, with time spent around those whose interests the humans are tasked to represent. A well-formulated option for this process is a Council of All Beings, which was conceptualized by Joanna Macy and others.[56] The Council of All Beings is a process designed to help participants step aside from their human identity and speak on behalf of another life-form. The Council of All Beings method originated among a group of people wishing to develop an emotionally richer response to Earth's destruction and to better understand both their place in the ecosphere as humans and how they should behave.

Their approach is supported by studies that show the power of activating our imagination to help build empathy for others. Taking the perspective of others can occur through storytelling, role play, and meditation practices designed to foster connection with other lifeforms and to somatically feel others' experiences. When we cognitively or physically take the perspective of others, we practice empathy, which then can increase our concern for their wellbeing.

Another option Gray and Curry suggest to help with creating proxies for non-human nature may lie in video games. Alf Seegert, the creator of a course titled, 'Video Games and Storytelling' at the University of Utah, has argued that their potential extends to evoking empathy for other species. A game that he cites as being especially effective in this regard is *Shelter* … in which the players take on the role of a badger mother who is trying to keep her cubs safe during a hazardous journey.[57] Gaming such as this can serve broadly as training for taking a non-human perspective, and thus it would not be necessary to develop different games for each context where ecodemocracy could be applied. The number of human votes and those given to proxies for non-human entities (and how those given to non-human nature are themselves divided) is a critical factor. A similar process of appointing two guardians to act on behalf of the Whanganui river in New Zealand occurred in 2017, when, as mentioned, the river was given the status of personhood, a decision that environmental lawmakers see as a paradigm shift from the colonial to the Maori worldview.[58]

Implications for the Treatment of Other Animals: Animal Lives Matter

In simple terms, a lot is at stake for all life forms on this planet, and their survival and wellbeing now all depend on the subsequent choices we make. Nonhuman animals matter to themselves, their social groups, populations and species, and their ecosystems. As sentient beings, nonhuman animals have an interest in the outcomes of their lives—nobody wants to die, nobody wants to suffer.

There is no doubt that we have the profound capacity to understand what another feels, which allows us to be compassionate and thus to avoid causing pain or suffering, and to act with an intention to do good for others. Concern for nonhuman animals must proceed from a theoretical grounding in ethics as the basis for practical action for animals. It is the animal rights movement that addresses animal suffering with the goal of animal liberation—decisively ending the exploitation of nonhuman animals altogether, which many scholars have claimed, will also lead to the liberation of ourselves and nature as a whole. This means that a 'piecemeal' and fragmented approach where we are satisfied with 'animal welfare' and giving chickens and pigs a few more inches of space in their cages before we send them a slaughterhouse, is not enough.

A goal of animal welfare will not free animals from their cages or save wild animals from hunters' bullets, anglers' hooks, or trappers' snares and other trap devices. We should not rest at asking for 'better protection' for wild animals in the form of prohibiting only certain types of hunting or compromising and accepting trapping 'only in certain situations.' We are not asking for better welfare or just better conditions for other human groups, we are demanding total equality in other social justice movements, which have been advanced through the protection of rights only. We must achieve the inclusion of animal rights in social justice movements and in existing social justice movements that continue to prioritize human rights while excluding nonhuman rights. For example, the Black Lives Matter, LGBTQI+, and the Women's' rights movements should become truly inclusive themselves by including other vulnerable animal victims and their rights.

Such an alliance of politics will garner more momentum so that it may successfully disrupt the power of white, male-hegemonic and capitalist corporations that not only have institutionalized animal torture but are currently destroying the planet. If we are serious about liberation efforts of any kind, then the Animal Industrial Complex—along with its counterpart of the Conservation Industrial Complex—must be dismantled. Sociologist Richard Twine's description of the Animal Industrial Complex, which also applies to the Conservation Industrial Complex as a

> Partially opaque network of relations between governments, public and private science, and the corporate (agricultural) conservation sector. Within the three nodes of the complex are multiple intersecting levels and it is sustained by an ideology that naturalizes the human as a consumer of other animals. It encompasses an extraordinary wide range of practices, technologies, identities and markets.[59]

The complexity here requires us to become acutely aware of our belief system as the very foundation of this complex interplay of mutually enforcing systems of animal exploitation. Repeated throughout this book, is the root of it all is our belief that we are separate from other animals and nature and superior to both. Our beliefs drive our actions and thus, what kind of food we buy, how we clothe ourselves, and the type of cosmetics and medicines we use, generally have a significantly detrimental impact on sentient animals. The consequences of our actions cost animals their lives. Some price is also paid by humans who, in many cases, are undocumented immigrants or migrant workers who are used to kill animals in slaughterhouses and meat processing plants, and who are being exploited (e.g., they do not get health insurance, even less than minimum wages). And we cannot forget the price paid by a degraded environment, being polluted and destroyed, which affects the animals and plants trying to survive, and eventually affects humans' physical and mental health as well.

The intersectionality of violence addressed in the previous chapter, and the various systems of oppression are made possible by the same human mindset, also overlap, intersect different parts of society and thus strengthen each other that way. Thus, we need to confront them all not separately but at once and at the same time. Part of this recognition entails the urgent need to move the issue of animal ethics and animal exploitation out of the impotent realm of personal choice into the potent realm of political action.

This is especially needed because the former (private sphere) is no match for the powerful Animal/Conservation Industrial Complex. Public policies, laws, and their enforcement reflect the dominant culture's view that animals exist to serve humans. Profound lifestyle changes and political activism for animals are needed to match and overcome the domination of the Animal/Conservation-Industrial Complexes, which continue to hold the power, influence, and control over public policy and thus the control over the public debate of the moral and legal status of other animals.

The political power of the Animal/Conservation Industrial Complex in maintaining the status quo to protect their profits through politics and lobbying cannot be underestimated. For example, environmental activists in third-world countries often 'disappear' or die (or both) while trying to protect nature from the destruction caused by corporations. US Congress has passed the Animal Enterprise Protection Act in 1992 and the Animal Enterprise Terrorism Act in 2006, to protect the interests of large-scale factory farms, large-scale fur operations, game farms and laboratories and their economically dependent businesses. These include animal breeders and suppliers of animals to these horrific institutions as well as cage manufacturers. "The enormous profits made from animal exploitation are zealously protected by existing relationships with governments and their regulatory mechanisms."[60] The same has occurred in terms of wildlife management and hunting and trapping, where aggressive actions by sportsmen's organizations led to the passing of Hunter Harassment Laws, to protect hunters from any interference of hunting (and trapping) by the "anti-hunting" public. A tidal wave of coordinated efforts by wealthy and influential (trophy) hunting organizations have swept across the United States, succeeding in protecting the right to fish and hunt in numerous states.

A commitment to the animal rights movements envisions the securing of moral rights for nonhuman animals and for them to become encoded into law as legal rights with meaningful and effective enforcement. Thus, we must challenge the laws and policies that allow wild animals to be killed by sportsmen in the first place (along with efforts to challenge any law and policy that directs the use of animals for human interests).

The animal rights movement faces a couple of prominent obstacles that also distinguishes it from other social justice movements. First, often social justice movements are spurred and supported by those whose self-interests are at stake—their activists are the agents of the change they themselves seek and need to benefit from, whereas humans take on the role to speak up for members of other species, whose voices are disregarded in a human-dominated society. Thus, it can be said that animal rights activists seek benefits for others without expecting benefits for themselves. Second, because humans are usually the beneficiaries of social justice movements, animal rights activists have the extended burden of working against the view held by people that they would have to "give up" something they enjoy (the steak, barbeque, hunting, trapping, and so on). In the end though, what matters in our lives is the way we treat others.

We have an enormous power to change animals' lives. A commitment to a non-violent relationship with other animals—that is the meaning and purpose of the vegan movement—is a good start to set an opposing wave in motion. And here the good news is that vegan products (plant-based) and cruelty-free products are increasingly available and represent a growing, economic force. However, they continue to be 'emerging' markets when what nonhuman animals need is a rapid and decisive change in human behavior. We all need to ask ourselves a crucial question: What does it take for me to commit to a life without exploitation of other beings?

Nonhuman animals cannot mobilize for themselves, so we have to on their behalf with a courageous heart, a decisive mind, and actions that are committed to achieving institutional change. For policies to change for wild animals, we must dismantle the system maintained by state wildlife agencies and sportsmen. Our goal must be that wildlife-related governmental agencies adopt Compassionate Conservation and work on behalf of wild animals and their habitats and ecosystems. Wild Lives Matter.

Concluding Remarks

Throughout this book, I have sought to show how the anthropocentric worldview underlying the North American Model of Wildlife Conservation causes suffering and death to hundreds of millions of wild animals every year. Fully conscious, competent and autonomous beings are treated like things without emotions, thoughts, and relationships. The disconnect between human and nonhuman animals has resulted in profound impacts on the planet that cannot be overestimated, particularly with respect to the violence against nonhuman animals.

Taken in total, this book's analysis illuminates an urgent need for a new ethical framework that serves humans and nonhumans alike without making one suffer for the other. The longstanding concept of species continuity is mirrored in varied facets, forms, and fashions, providing for species diversity and individual animal subjectivity, humans and animals. Suffering impacts individuals; from there, it reverberates through their families and comrades until it affects an entire population and even the species. Flourishing must also start with the individual and her/his relationships with others. While other animals have their own species-specific relationships among one another, they also depend on the kind of relationships we choose to have with them. Thus, healthy relationships are those formed by humans toward nonhumans that cause their flourishing, not destruction and forced death. It is time for us to think and feel beyond ourselves and across the species lines. We know what to do, let's do it.

What You Can Do

- You can make a profound difference for all animals, domestic and wild, and for the health of the planet more broadly, by switching, even slowly, to a vegan diet. There is an overlap between billions of sentient and defenseless nonhuman animals incarcerated by industries, brutally treated and slaughtered for food, and wolves and other predators killed for 'livestock' protection.

- Visit and support wildlife and farm animal sanctuaries, perhaps start one yourself.
- Become an animal rights attorney.
- Support GENIE, the Global Ecocentric Network for Implementing Ecodemocracy: (20+) Global Ecocentric Network for Implementing Ecodemocracy (GENIE) | Facebook
- Become an ambassador of the Nature Needs Half initiative: Home - Nature Needs Half
- Get involved in your country's or state's legislature, where all too often bills detrimental to wild animals are being passed.
- Consider running for office to promote wildlife- and Earth-friendly policies and vote for legislators who champion these.
- Bring attention to the plight of wild animals by writing letters to your local newspaper, and on social media.
- If you belong to a congregation, the peace movement, or any social justice, political or conservation organization, question them on their stance on wildlife individuals (you'll be surprised).
- You can also join an Animal Rights organization, Earth rights group or a Community Environmental Legal Defense Fund chapter or support the Earth Law Center.
- If you are a student of wildlife conservation, biology etc., inquire about compassionate conservation.
- Volunteer at your local humane society or animal shelter.
- You can explore your own attitudes toward nonhuman animals and consider how you view them – for example:[61]

  – when you see an animal, do you see them as food or as a sentient being and deserving your respect and care?
  – do you believe that your status as a human being entitles you to participate in their suffering and death?
  – if you are a 'meat' eater, do you ever think about who these nonhumans had been and how they suffered and died?

- are your beliefs and your lifestyle harming nonhuman animals, or do they add more compassion to the world?
- where do your beliefs originate? When did you hear them for the first time and become to believe them? Who told you these?
- would it benefit you and others to change your beliefs and attitudes and actions toward nonhuman animals? How would you go about it? Do you need support? Attitudes and actions of kindness, respect and compassion expand and reinforce each other just as systems of violence and oppression.

Notes

1. Personal communication with Paul N. in December 2020.
2. Aaltola *Animal Suffering*, 158 (a reference to Mary Midgley).
3. For more details on tools such as argument analysis and conflict resolution to help guide appropriate decision-making, see Nelson et al. "Cecil."
4. Panksepp, "Affective Consciousness."
5. Batavia et al., "Emotion."
6. Young et al., "Empathy."
7. Bekoff and Pierce, *Wild Justice*, 94.
8. Ibid, 88.
9. Ibid, 90.
10. Ibid, 94.
11. Ibid, 94.

    But it is also worth noting that our current status of recognizing empathy in other animals may have to do more with our limited abilities than anything else.
12. Andersen, "Journey."
13. Bekoff and Pierce, *Wild Justice,* 87.
14. Nurse and Wyatt, *Wildlife Criminology.*
15. Kymlicka and Donaldson (2014, p. 120).
16. Luke, "Taming," pp. 303–310.
17. McShane, "Anthropocentrism."
18. Batavia and Nelson, "Heroes or Thieves?"
19. Donovan, "Feminism," 323–324.
20. Kynlick and Donaldons (2014, p. 117).

21. Young et al. (2018, p. 334).
22. Ibid, p. 334.
23. WWF, "Living Planet Index."
24. Dean, "Key Findings from IPBES Report."
25. Crist "Great Reckoning."
26. WWF, "Losing Nature?"
27. Ripple et al., "Eating."
28. González et al. "Meat Consumption."
29. Crist et al. "Protecting Half the Planet."
30. Ramp and Bekoff, "Compassion."
31. Wallach et al. "Recognizing Animal Personhood," 6.
32. Ramp and Bekoff, "Compassion."
33. Wallach et al., "Summoning."
34. Ibid., 1.
35. Bekoff, "Conservation Lacks Compassion."
36. Hervieux et al. "Managing Wolves" (also, see the response by Proulx et al., "Poisoning," strongly condemning the study as well as its publication).
37. Foran, *Subjugation,* 219–220.
38. Baker, "Translocation Biology."
39. Ramp and Bekoff (2015), "Compassion."
40. Wallach et al., "Summoning."
41. Ibid, 7.
42. Bekoff, "First Do No Harm," 25.
43. Publications with critical views of CC include: Hayward et al., "Deconstructing"; Oommen et al. "Fatal Flaws"; Griffin et al. "Clashes."
44. Donaldson and Kymlicka, *Zoopolis*, 176.
45. Taylor, "Ethics of Respect," 202, with original emphasis.
46. Fox, "Coyotes And Humans."
47. Nonhuman Rights Project, "Recognition and Rights."
48. Cullinan, *Wild Law*, 108.
49. Heister, "Beyond the North American Model."
50. Donaldson and Kymlicka, *Zoopolis,* 176.
51. Rincoast Conservation Foundation, "Tenure."
52. Hadley, "Nonhuman Animal Property," 476, 473.
53. Bradshaw, *Wildlife as Property Owners*, 3.
54. MacKinnon, "Of Mice and Men," 263, 270.
55. Plumwood, "Democracy and Failed Ecology," 142.

56. Work That Reconnects Network, "Council of All Beings."
57. Seegert, "Pixels and Pathos."
58. Hollingsworth/CNN, "River as a Legal Person."
59. Twine, "Revealing." He uses brackets around the agricultural because the definition is not limited to this sector and just indicates his interest for this article.
60. Stallwood, "Political Turn for Animals."
61. Spanjol, "Wild Crimes."

References

Aaltola, Elisa. *Animal Suffering: Philosophy and Culture.* United Kingdom: Palgrave Macmillan, 2012.

Adams, Carol. "The War on Compassion," In *The Feminist Care Tradition in Animal Ethics,* eds. Josephine Donovan and Carol J. Adams., pp. 21-36, 2007.

Adams, Carol J. "The Feminist Traffic in Animals." In *Ecofeminism: Women, Animals, Nature,* edited by Greta Gaard, 195–218. Philadelphia: Temple University Press, 1993.

Adams, Carol J. *The Sexual Politics of Meat: A Feminist vegetarian Critical Theory.* New York: Continuum International Publishing Group, 1990.

Adams, Carol J., and Lori Gruen, eds. *Ecofeminism: Feminist Intersections with Other Animals and the Earth.* New York and London: Bloomsbury Academic, 2014.

"Agency Capture." Oxford Dictionary. Agency capture - Oxford Reference.

Almiron, Núria and Fernández, Laura. Including the Animal Standpoint in Critical Public Relations Research. *Public Relations Inquiry,* 10(2) (2021): 139–161 (accepted manuscript). The full paper is available at: https://journals.sagepub.com/doi/10.1177/2046147X211005368

American Game Protective Association. "A Proposed American Game Policy to be Discussed at the Seventeenth Annual American Game Conference, New York City, December 1-2, 1930. Jaques and Company, accessed January

A. Heister, *Beyond the North American Model of Wildlife Conservation,* The Palgrave Macmillan Animal Ethics Series, https://doi.org/10.1007/978-3-031-14149-2

22, 2022, A Proposed American Game Policy to be Discussed at the Seventeenth Annual American Game Conference, New York City, December 1-2, 1930 - Report to the National Game Conference on American Game Policy | Adirondack Experience (pastperfectonline.com).

Andersen, Ross. "Journey into the Animal Mind: What science can tell us about how other creatures experience the world." The Atlantic, March 2019. Do Animals Have Feelings? - The Atlantic.

"Animal Legal Defense Fund, Inc. v. Department of Environmental Conservation, 123 A.D.2d 368 (1986)" Animal Legal Defense Fund, Inc. v. Department of Environmental Conservation, 123 A.D.2d 368 (1986).

AnimalsAsia. "Animals Asia Brings 101 Bears Home in Largest Ever Operation of its Kind." Animals Asia brings 101 bears home in largest ever operation of its kind.

Animal Welfare Institute. "Wildlife Killing Contests." Accessed November 19, 2021. Wildlife Killing Contests | Animal Welfare Institute (awionline.org).

Ashe, Daniel M. "Cruelty toward wolves is erasing conservation efforts. It's time to reinstate their protections." August 3, 2021. Opinion | Cruelty toward wolves is erasing conservation efforts. It's time to reinstate their protections. - The Washington Post.

Association of Fish and Wildlife Agencies (AFWA). National Fur Harvest Database. Season_1970-2018_AFWA_Fur_Harvest.xlsx (live.com).

Associated Press, "7 Collared Yellowstone Park Wolves Shot by Hunters," Missoulian, November 15, 2012, modified October 3, 2014. 7 collared Yellowstone Park wolves shot by hunters | State & Regional | missoulian.com.

Associated Press. "6 People Cited after Hunters Kill About 50 Elk in Montana." October 28, 2020. 6 people cited after hunters kill about 50 elk in Montana - Washington Times.

Associated Press, "Montana considers trapping restrictions Near Yellowstone, Glacier Parks; trapper group opposed." Casper Star Tribune. May16, 2015. www.trib.com.

Associated Press. "Beaver trap breaks man's leg in Oregon," Herald Net, December 19, 2013, Beaver trap breaks man's leg in Oregon | HeraldNet.com.

Baker, Liv. "Translocation Biology and the Clear Case for Compassionate Conservation." Israel Journal of Ecology and Evolution, Vol. *63*, no. 3-4, (2017): 52-60. doi: https://doi.org/10.1163/22244662-20181026.

Batavia, Chelsea, Michael P. Nelson, Jeremy T. Bruskotter, Megan S. Jones, Esty Yanco, Daniel Ramp, Marc Bekoff and Arian D. Wallach. "Emotion as a

Source of Moral Understanding in Conservation." *Conservation Biology*, Vol. 0 no. 0 (2020): 1-8.

Batavia, Chelsea and Michael P. Nelson. "Heroes of Thieves? The Ethical Grounds for Lingering Concerns About New Conservation." J Environ Stud Sci 7, 394-402 (2017). https://doi.org/10.1007/s13412-016-0399-0.

Bekoff, Marc. "What Can Dogs, Coyotes, and Opossums Tell Us About Death?" October, 3, 2021. What Can Dogs, Coyotes, and Opossums Tell Us about Death? | Psychology Today.

Bekoff, Marc, ed. *Ignoring Nature No More: The Case for Compassionate Conservation*. Chicago: University of Chicago Press, 2013.

Bekoff, Marc. "Animal Minds and the Foible of Human Exceptionalism." *The Huffington Post* (October 5, 2011). Retrieved from http://www.huffingtonpost.com/marc-bekoff/animal-minds-and-the-foib_b_919028.html.

Bekoff, Marc. "Conservation Lacks Compassion." *New Scientist* 207 (2010a): 24–25. doi:https://doi.org/10.1016/s0262-4079(10)62080-5.

Bekoff, Marc. First Do No Harm. New Scientist, Opinion. August 28, 2010b. First do no harm [146].pdf.

Bekoff, Marc. *The Animal Manifesto: Six Reasons for Expanding Our Compassion Footprint. Novato, CA:* New World Library, 2010c.

Bekoff, Marc, and Jessica Pierce. *Wild Justice: The Moral Lives of Animals*. United States: University of Chicago Press, 2009.

Bekoff, Marc. *The Emotional Lives of Animals*. United States: New World Library, 2007.

Bekoff, Marc. *Minding Animals: Awareness, Emotions, and Heart*. United States: Oxford University Press, 2002.

Best, Stephen. *The Politics of Total Liberation: Revolution for the 21st Century*. New York: Palgrave Macmillan, 2014.

Beston Henry. *The Outermost House: A Year of Life On The Great Beach of Cape Cod*. New York: Henry Holt and Company, 1928.

Bever, Lindsey. "A mother Bear and Two 'Shrieking' Newborn Cubs were Killed in their Den by Hunters, Troopers say." August 10, 2018. Hunters Andrew and Owen Renner charged after killing mother bear and newborn cubs - The Washington Post.

Bixby, Kevin. "Why Hunting Isn't Conservation, and Why It Matters," 2020. Why Hunting Isn't Conservation, and Why It Matters - Rewilding.

Bolen, Eric G., and William L. Robinson. *Wildlife Ecology and Management, Fifth Edition*. Upper Saddle River, NJ: Prentice Hall, 2003.

Boone and Crockett Club. "North American Model of Wildlife Conservation." Accessed May 1, 2021. Boone and Crockett Club | (boone-crockett.org).

Boone and Crockett Club. "BC&C Welcomes the NRA to its 30th Big Game Awards." News, May 30, 2019. www.boone-crockett.org.

Boone and Crockett Club. "History of the BCC". History of the Boone and Crockett Club | Boone and Crockett Club (boone-crockett.org).

Boone and Crockett Club, "Action: What's Next." Boone and Crockett Club/ American Wildlife Conservation Partners. www.boone-crockett.org.

Boone and Crockett Club (BCC). "Mission." About the Boone and Crockett Club | Boone and Crockett Club (boone-crockett.org).

"Boone and Crockett Club 30th Awards Auction." Online Hunting Auctions.com.

Born Free USA Calls for an End to Trapping on National Wildlife Refuges. September 15, 2020. Born Free USA Calls for an End to Trapping on National Wildlife Refuges (prnewswire.com).

Born Free USA. "Poll: Americans Strongly Oppose Trapping on National Wildlife Refuges." Trapping. Last modified July 18, 1999. Retrieved January 19, 2022, from Poll: Americans Strongly Oppose Trapping on National Wildlife Refuges | Born Free USA.

Bradshaw, Karen. *Wildlife as Property Owners: A New Conception of Animal Rights.* London: The University of Chicago Press, 2020.

Brands, H. W. *T.R.: The Last Romantic* (6th ed.). New York, NY: Basic Books, 1997.

Brooks, Talasi. "Slaughtered Pups and Maimed Wolves in Idaho Demonstrate the Effects of Federal Delisting." Accessed September 3, 2020. https://www.westernwatersheds.org/2020/09/slaughtered-pups-and-maimed-wolves-in-idaho-demonstrate-the-effects-of-federal-delisting/.

Brown, Kelsey, "A Feminist Analysis of Human and Animal Oppression: Intersectionality Among Species" (2016). Scholars Week. 2. https://cedar.wwu.edu/scholwk/Departmental_Presentations/Women_Gender_Sexuality_Studies/2.

Bull, Bartle. "Hunter-Conservationist or… Jekyll & Hyde?" *Time Magazine*, July 31 2015.

Carrington, Damien. "Eating Less Meat Essential to Curb Climate Change," Our World, United Nations University, 2014. https://ourworld.unu.edu/en/eating-less-meat-essential-to-curb-climate-change-says-report.

Center for Biological Diversity. "115 Top U.S. Wolf Experts, Scientists Urge Biden Administration to Restore Federal Protections for Gray Wolves: *State*

Wildlife Agencies Reject Science, Demonstrate Inability to Sustain Wolf Populations." May 13, 2021. 115 Top U.S. Wolf Experts, Scientists Urge Biden Administration to Restore Federal Protections for Gray Wolves - Center for Biological Diversity.

Center for Biological Diversity. "Reward Increased to $10,000 for Info on Killing of Endangered Washington Wolf." June 4, 2019. https://biological-diversity.org/w/news/press-releases/reward-increased-10000-info-killing-endangered-washington-wolf-2019-06-04/.

Center for Biological Diversity. "Listing Species Under the Endangered Species Act." https://www.biologicaldiversity.org/programs/biodiversity/endangered_species_act/listing_species_under_the_endangered_species_act/index.html.

Chadwick, Douglas H. *The Wolverine Way*. United States: Patagonia Books, 2010.

Chaney, Rob. "Family Pulls Award Over Rocky Mountain Elk Foundation's Wolf Policy." Updated October 3, 2014. http://missoulian.com/news/local/family-pulls-award-over-rocky-mountain-elk-foundation-s-wolf/article_e42c3c42-d143-11e1-acf9-001a4bcf887a.html.

Clark, Susan G. and Christina Milloy, "The North American Model of Wildlife Conservation: An Analysis of Challenges and Adaptive Options." In *Large Carnivore Conservation: Integrating Science and Policy in the North American West*, edited by Susan Clark and Murray Rutherford, Chicago: University of Chicago Press, 2014. DOI: https://doi.org/10.7208/Chicago/9780226107547.001.0001

Cohn, Priscilla and Andrew Linzey. "Hunting as a Morally Suspect Activity," In: *The Link between Animal Abuse and Human Violence*, edited by A. Linzey, pp. 317-28. Eastbourne: Sussex Academic Press, 2009.

Cole, Ken. "The Carnage of Wolf Trapping in Idaho. New Public Records Request Shows High Non-Target Capture and Mortality." The Wildlife News, April 13, 2020. The Carnage of Wolf Trapping in Idaho. New Public Records Request Shows High Non-Target Capture and Mortality. | The Wildlife News.

Cole, Ken. "State Public Records Request Shows Widespread Capture and Mortality of Non-target Animals Related to Idaho Wolf Trapping During 2011/2012 Trapping Season." *The Wildlife News*. (February 14, 2013). Retrieved from The Wildlife News, http://www.thewildlifenews.com/2013/02/14/state-public-records-request-shows-widespread-capture-and-mortality-of-non-target-animals-related-to-idaho-wolf-trapping-during-20112012-trapping-season/.

Congressional Sportsmen's Foundation (CSF). "About." Congressional Sportsmen's Foundation. Retrieved August 12, 2021, from https://congressionalsportsmen.org/about.

Conservation Force. "Ranching for Restoration," Pamphlet. www.conservationforce.org. Available at: ranch (conservationforce.org).

Conservation Force. "Will Lion Hunting Survive? And More …" internationalhunting articles | conservation-force (conservationforce.org).

Conservation Leaders for Tomorrow. "About Conservation Leaders for Tomorrow." About Conservation Leaders for Tomorrow | Conservation Leaders for Tomorrow (clft.org).

Conservation Visions, and The Rocky Mountain Elk Foundation. *Opportunity for All: The Story of the North American Model of Wildlife Conservation* [DVD]. United States, Rocky Mountain Elk Foundation, 2012.

Corbett, Jessica. "Reckless, Violent, Massacre of 570 Wolves and Wolf Pups in Idaho Bolsters Alarm Over Trump Attack on Species Protections." Accessed December 1, 2020. 'Reckless, Violent, Massacre' of 570 Wolves and Wolf Pups in Idaho Bolsters Alarm Over Trump Attack on Species Protections (commondreams.org).

Crist, Eileen. "Our Great Reckoning: Eileen Crist On The Consequences Of Human Plunder." Interview by Leath Tonino. The Sun Magazine, December 2020. Our Great Reckoning | By Leath Tonino | Issue 540 | The Sun Magazine.

Crist, Eileen. *Abundant Earth: Toward an Ecological Civilization. Chicago Scholarship Online*, September 2019a. DOI: https://doi.org/10.7208/Chicago/9780226596945.001.0001.

Crist, Eileen, 2019b, interview with Marc Bekoff. "Abundant Earth: Offers a positive Vision for the Future." April 10, 2019. "Abundant Earth" Offers a Positive Vision for the Future | Psychology Today.

Crist Eileen. "The Affliction of Human Supremacy." *The Ecological Citizen* 1 (2017): 61–64.

Crist, Eileen "The Inner Life of Earthworms: Darwin's Argument and its Implications." In *The Cognitive Animal*, edited by Marc Bekoff, Colin Allen, and Gordon Burghardt. MIT Press, 2002.

Crist, Eileen. "Abundant Earth: Toward an Ecological Civilization," DOI: https://doi.org/10.7208/Chicago/9780226596945.001.0001

Crist, Eileen, and Helen Kopnina. "Unsettling Anthropocentrism." *Dialect Anthropology*, Vol. 38, (November 21, 2014): 387–396. doi:10.1007/s10624-014-9362-1.

Crist, Eileen, Helen Kopnina, Philip Cafaro, Joe Gray, William J. Ripple, Carl Safina, John Davis, Dominick A. DellaSala, Reed F. Noss, Haydn Washington, Holmes Rolston III, Bron Taylor, Ewa H. Orlikowska, Anja Heister, William S. Lynn and John J. Piccolo. "Protecting Half the Planet and Transforming Human Systems Are Complementary Goals." Frontiers in Conservation Science, 18 November 2021 | https://doi.org/10.3389/fcosc.2021.761292.

Cullinan, Cormac. Wild Law: A Manifesto for Earth Justice. Green Books, The Gaia Foundation, London: United Kingdom.

Daigle, J., Muth, R., Zwick, R., and Glass, R. (1998). Sociocultural Dimensions of Trapping: A Factor Analytic Study of Trappers in Six Northeastern States. *Wildlife Society Bulletin*, *26*(3), 614–625.

Damasio, Antonio and Kaspar Meyer. "Behind the Looking Glass." *Nature* 454 (2008).

Darimont, Chris T., Fox, Caroline H., Bryan, Heather M., and Thomas E. Reimchen. "The Unique Ecology of Human Predators." *Science* 349, no. 6250 (Aug 21, 2015): 858-860. DOI: https://doi.org/10.1126/science.aac4249.

Darwin, Charles. "Vermin and Traps." *Gardeners' Chronicle and Agricultural Gazette* no. 35 (29 August, 1863): 821-822. Reprinted by Defenders of Wildlife, 1980.

Darwin, Charles. *The Formation of Vegetable Mould Through the Action of Worms, with Observations on their Habitats.* Chicago: University of Chicago Press, 1881/1986.

Davis, Karen. "An Open Letter to the Guardian." August 20, 2020, a Response to: From Farm to Factory: The Unstoppable Rise of American Chicken, The Guardian, Aug. 17, 2020. An Open Letter to The Guardian - 20 August 2020 - UPC (upc-online.org).

Dean/United Nations Foundation. "Key Findings to Know from the IPBES Report on Biodiversity." May 6, 2019. Key Findings to Know from the IPBES Report on Biodiversity | unfoundation.org |

Deckha, Maneesha. "Critical Animal Studies and Animal Law." *Animal Law Review, Vol 18,* (2012): 207-236.

de Waal, Frans. *The Age of Empathy: Nature's Lessons for a Kinder Society.* Toronto: McClelland and Stewart, 2009.

Donaldson, Sue and Will Kymlicka. *Zoopolis: A Political Theory of Animal Rights.* Oxford: Oxford University Press, 2011.

Donovan, Josephine. Aesthecizing Animal Cruelty. *College Literature*, Vol 38, No. 4 (Fall 2011): 202-217.

Donovan, Josephine "Animal Rights and Feminist Theory." In *Ecofeminism: Women, Animals, Nature,* edited by Greta Gaard, 167-194. Philadelphia: Temple University Press, 1993.

Donovan, Josephine and Carol J. Adams. "Introduction." In: *The Feminist Care Tradition in Animal Ethics: A Reader*, edited by Josephine Donovan and Carol Adams, 1–15. New York, NY: Columbia University Press, 2007.

Donovan, Josephine (originally in 2006), *Caring to Dialogue: Feminism and the Treatment of Animals In: The Feminist Care Tradition in Animal Ethics*, Eds. Josephine Donovan and Carol J. Adams. New York: Columbia University Press, 2007.

Duda, M. D., T. Beppler, and J. Organ. 2017. "The Growth of Sport Shooting Participation: what does this Trend mean for Conservation Revenue? The Wildlife Professional 11 (2), (2017): 38-41.

Ehrlick, Darrell. "Believe Gianforte: He is a man of his actions: Wolf-trapping governor continues to demonstrate what kind of leader he will be." Accessed April 1, 2021. Believe Gianforte: He is a man of his actions – Daily Montanan.

Elliot, Will. "Mahoney Advises Hunters to Connect Past with Future." *Buffalo News*, June 27, 1999. https://buffalonews.com/news/mahoney-advises-hunters-to-connect-past-with-future/article_791e8e2b-c63f-5202-aed9-cbf555010f76.html.

Esteban, Joan-María, and Debraj Ray. "On the Measurement of Polarization." *Econometrica,* Vol. 62, no. 4 (1994): 819-51. doi:https://doi.org/10.2307/2951734.

Faunalytics. "Henry Bergh, The Unlikely Champion Of Animals." July 10, 2014. Henry Bergh, The Unlikely Champion Of Animals - Faunalytics.

Feldpausch-Parker, Andrea, Israel Parker, and Elizabeth Vidon. "Privileging Consumptive Use: A Critique of Ideology, Power, and Discourse in the North American Model of Wildlife Conservation." *Conservation and Society*. Vol. 15, No. 1 (2017): 33-40.

Food and Agricultural Organization of the United Nations, FAO 2021 Available at: FAOSTAT: Accessed on 2 December 2021.

Foran, Max. *The Subjugation of Canadian Wildlife: Failures of Principle and Policy*. Montreal & Kingston: McGill-Queen's University Press, 2018.

Foundation for Wildlife Management. "Tired of Wolves Interrupting your Hunt and Displacing Our Wildlife?" Accessed May 1, 2020. https://www.foundationforwildlifemanagement.org/

Fox, Camilla H. "Coyotes and Humans: Can We Coexist?" In Proceedings of the Annual Vertebrate Pest Conference, edited by RM Timm and JM O'Brien, pp. 287-293. Davis: University of California Davis.

Fox, Camilla, and Christopher M. Papouchis. *Cull of the Wild: A Contemporary Analysis of Wildlife Trapping in the United States*. United States: Animal Protective Institute, 2004.

Francione, Gary L. *Introduction to Animal Rights: Your Child or Your Dog?* Philadelphia: Temple University Press, 2000.

Frankfurt School and Critical Theory | Internet Encyclopedia of Philosophy (utm.edu).

Fritzen, Gabe. "Anahareo: A Portrait of a Famous Canadian Woman." 2002. About Her Name - Anahareo.

"Frivolous." Merriam-Webster Dictionary Online. Frivolous Definition and Meaning - Merriam-Webster.

Funk, M., and Julie Schuppel. "Strangulation Injuries." *Wisconsin Medical Journal*, Vol. 102 no. 3 (2003): 41-45.

Gaard, Greta, ed. *Ecofeminism: Women, Animals, Nature*. Philadelphia: Temple University Press, 1993.

Geist, Valerius. "The North American Model of Wildlife Conservation: A Means of Creating Wealth and Protecting Public Health While Generating Biodiversity." In *Gaining Ground: In Pursuit of Ecological Sustainability*, edited by David Lavigne. Ireland: International Fund for Animal Welfare, 2006.

Geist, Valerius, Shane P. Mahoney, and John F. Organ. "Why Hunting Has Defined the North American Model of Wildlife Conservation." *66th Transactions of The North American Wildlife and Natural Resources Conference*, 2001. Abstract retrieved from http://conservationvisions.com/sites/default/files/why_hunting_has_defined_the_north_american_model_of_wildlife_conservation.pdf.

Gilbert, Frederick, and Norman Gofton. "Terminal Dives in Mink, Muskrat and Beaver. *Physiology and Behavior,*" Vol. 28 no. 5 (1982): 835–840. doi: 10.1016/0031-9384(82)90200-1.

Gill, R. Bruce. 2004. "Challenges of Change: Natural Resource Management Professionals Engage Their Future." Pages, pp. 35- 46, In: M. J. Manfredo, J. J.Vaske, B. L. Bruyere, D. R. Field, and P. J. Brown, editors. *Society and natural resources: a summary of knowledge*, Modern Litho, Jefferson City, Missouri, USA.

Gilligan, Carol. *In a Different Voice: Psychological Theory and Women's Development*. Cambridge, Massachusetts: Harvard University Press, 1982.

Gonçalves, Eduardo. "The Scale and Power of the Trophy Hunting Industry Is Astonishing." Interview by Dylan Forest, Animal People Forum, February 7, 2021. "The Scale and Power of the Trophy Hunting Industry Is Astonishing": An Interview With Eduardo Gonçalves (animalpeopleforum.org).

Gonçalves, Eduardo. *Trophy Hunters Exposed: Inside the Big Game Industry.* London: Green Future Books, Ltd., 2020. Kindle.

González, Neus, Montse Marquès, Martí Nadal, and José L. Domingo. "Meat consumption: Which are the current global risks? A review of recent (2010–2020) evidences." *Food Res Int.* 137: 109341. (May 2020): doi: https://doi.org/10.1016/j.foodres.2020.109341.

Graça, J., Calheiros, M. M., Oliveira, A., & Milfont, T. L. (2018). Why are Women less likely to Support animal exploitation than men? The mediating roles of social dominance orientation and empathy. *Personality and Individual Differences, 129,* 66–69. https://doi.org/10.1016/j.paid.2018.03.007.

Graham Jr., Frank. *Man's Dominion: The Story of Conservation in America.* Drawings by John Pimlott, New York: M. Evans & Co, 1971.

Griffin, Andrea S., Alex Callen, Kaya Klop-Toker, Robert J. Scanlon, and Matt W. Hayward. "Compassionate Conservation Clashes with Conservation Biology: Should Empathy, Compassion, and Deontological Moral Principles Drive Conservation Practice?" Front Psychol. (2020): 11:1139. doi: https://doi.org/10.3389/fpsyg.2020.01139. PMID: 32536896; PMCID: PMC7269110.

Gross, W.H. "Chip. National Trappers Association Celebrates Golden Anniversary." ESPN. July 31, 2009. National Trappers Association celebrates golden anniversary (espn.com).

Gruen, Lori. "Dismantling Oppression: An Analysis of the Connection Between Women and Animals." In *Ecofeminism: Women, Animals, Nature,* edited by Greta Gaard, 167-194. Philadelphia: Temple University Press, 1993.

Gullone, Eleonora. "To Minimize Animal Suffering, Broaden the Definition of Animal Cruelty." *Animal Sentience* 6, no. 7 (2016), DOI: https://doi.org/10.51291/2377-7478.1100.

Gullone, Eleonora. "Conceptualising Animal Abuse with an Antisocial Behaviour Framework." Animals (Basel). 2011 Mar; 1(1): 144–160. Published online 2011 Jan 26. doi: https://doi.org/10.3390/ani1010144.

Hadley, John. "Nonhuman Animal Property: Reconciling Environmentalism and Animal Rights." In: *Animal Rights,* edited by Clare Palmer, 465-475. Aldershot, UK: Ashgate, 2008.

Harmonist. "The Animal Industrial Complex." January 4, 2011. Available at: The Animal Industrial Complex | Harmonist.

Hayward, Matt W., Alex Callen, Benjamin L. Allen, Guy Ballard, Femke Broekhuis, Cassandra Bugir, Rohan H. Clarke, John Clulow, Simon Clulow, Jennifer C. Daltry, Harriet T. Davies-Mostert, Peter J. S. Fleming, Andrea S. Griffin, Lachlan G. Howell, Graham I. H. Kerley, Kaya Klop-Toker, Sarah Legge, Tom Major, Ninon Meyer, Robert A. Montgomery, Katherine Moseby, Daniel M. Parker, Stéphanie Périquet, John Read, Robert J. Scanlon, Rebecca Seeto, Craig Shuttleworth, Michael J. Somers, Cottrell T. Tamessar, Katherine Tuft, Rose Upton, Marcia Valenzuela-Molina, Adrian Wayne, Ryan R. Witt, Wolfgang Wüster. "Deconstructing Compassionate Conservation." The Society for Conservation Biology, Vol 33, no. 4, (August 2019): 760-768. https://doi.org/10.1111/cobi.13366.

Heinrich, Bernd. *Mind of the Raven: Investigations and Adventures with Wolf-Birds.* New York: Cliff Street Books, 1999.

Heister, Anja. Earth's Animal Nations: It's Time for Human Compassion," In *Visions for a Post-Covid World: Defining a Radically New Normal,* edited by Joe Gray, Dixi Books, 2021, Palmers Green, London, England, pp. 15-28.

Heister, Anja. "Beyond the North American Wildlife Conservation Model and Towards Earth Rights." The Ecological Citizen 3 (Suppl A), (2019): 67–74. v03sa-09.pdf (ecologicalcitizen.net).

Hendershott, Anne. "Empathy Deficit." The Spectator, April 2021. Confronting the Real Empathy Deficit | The American Spectator | USA News and Politics.

Hervieux, Dave, Mark Hebblewhite, Dave Stepnisky, Michelle Bacon, and Stan Boutin. *Canadian Journal of Zoology,* 19 November 2014. "Managing Wolves (Canis lupus) to Recover threatened woodland caribou (Rangifer tarandus caribou) in Alberta" (cdnsciencepub.com).

Hoffman, Austin D. "A Brief History of Wild Wolves." 2021. A Brief History of Wild Wolves | Mission: Wolf (missionwolf.org).

Holleman, Marybeth. "Wolf Populations Should be Assessed by Packs, Not Individuals." *High Country News.* November 7, 2013. http://www.hcn.org/wotr/the-governments-decision-to-delist-gray-wolves-is-flawed-because-it-counts-animals-instead-of-families.

Hollingsworth/CNN. "This River This river in New Zealand is legally a person. Here's how it happened." December 11, 2020. How New Zealand's Whanganui River is legally a person - CNN.

Holmes, Madelyn. *American Women Conservationists: Twelve Profiles.* Jefferson, NC: McFarland & Company, Inc. 2004.

Horner, Susan Morath. "Embryo, not Fossil: Breathing Life into the Public Trust in Wildlife. *Land and Water Law Review,* Vol. 35, Iss.1 (2000): 1–66.

Horsthemke, Kai. *Animal Rights Education.* The Palgrave Macmillan Animal Ethics Series, eds. Andrew Linzey, Priscilla N. Cohn and Clair Linzey. Oxford, UK: Oxford Centre for Animal Ethics, 2018.

Horta, Oscar. "Concern for Wild Animal Suffering and Environmental Ethics: What are the Limits of the Disagreement?" *Les Ateliers de l'Éthique / the Ethics Forum* 13 (1):85-100 (2018). DOI: https://doi.org/10.7202/1055119ar.

Hribal Jason. *Fear of the Animal Planet: The Hidden History of Animal Resistance.* California: AK Press/Counterpunsch, 2010.

Inman, Bob, K. Podruzny, T. Smucker, A. Nelson, M. Ross, N. Lance, T. Parks, D. Boyd and S. Wells. 2018. "Montana Gray Wolf Conservation and Management 2018 Annual Report." Montana Fish, Wildlife and Parks, 201 8-montana-wolf-annual-report-final-8.23.2019.pdf (mt.gov).

Inman, B., K. Podruzny, A. Nelson, D. Boyd, T. Parks, T. Smucker, M. Ross, N. Lance, W. Cole, M. Parks, and S. Wells. 2019. Montana Gray Wolf Conservation and Management 2019 Annual Report. Montana Fish, Wildlife and Parks. https://leg.mt.gov/content/Committees/Interim/2019-2020/EQC/Meetings/sept-2020/fwp-wolf-2020-report.pdf.

IAFWA (International Association of Fish and Wildlife Agencies). *Trapper Educational Manual.* 2005. Retrieved from http://www.fishwildlife.org/files/Trapping_Student_Manual_2005.pdf.

IFAW (International Fund for Animal Welfare). 2016, Executive Summary, pg. 4: Killing_For_Trophies.pdf. killing for trophies: an analysis of global trophy hunting trade | IFAW.

Institute for Critical Animal Studies. Institute for Critical Animal Studies (ICAS) | Education. Liberation. Transformation.

International Union for Conservation of Nature (IUCN). "Conservation efforts bring cautious hope for African rhinos - IUCN Red List." Last modified 19 March, 2020. Conservation efforts bring cautious hope for African rhinos - IUCN Red List | IUCN.

International Union for Conservation of Nature (IUCN). Commission on Environmental, Economic and Social Policy, n.d. "The North American Wild Harvest Initiative (WHI)." The North American Wild Harvest Initiative (WHI) | IUCN.

Iossa, Graziella, Soulsbury, C.D., and Stephen Harris. "Mammal Trapping: a Review of Animal Welfare Standards of Killing and Restraining Traps." *Animal Welfare,* Vol.16, no. 3 (August 2007): 35–352.

Jacobs, Julia. "Alaska Hunter Who Killed Cubs in Bear Den Gets 3 Months in Jail." New York Times. January 26, 2019. Alaska Hunter Who Killed Cubs in Bear Den Gets 3 Months in Jail - The New York Times (nytimes.com).

Jacobson, C. A., J. F. Organ, D. J. Decker, G. R. Batcheller, and L. Carpenter. A Conservation institution for the 21st century: Implications for state wildlife agencies. *Journal of Wildlife Management* 74, no. 2 (2010): 203–209.

Jane Fonda. "Ted Turner and Prairie Dogs." August 27, 2013. https://www.janefonda.com/2013/08/ted-turner-prairie-dogs/.

Joy, Melanie. *Powerarchy: Understanding the Psychology of Oppression for Social Transformation*. Oakland, California: Berrett-Koehler Publishers, Inc., 2019.

Kemmerer, Lisa. "Hunting Tradition: Treaties, law, and Subsistence Killing," *Animal Liberation Philosophy and Policy Journal. 2, no.* 2 (2004).

Kheel, Marti. *Nature Ethics: An Ecofeminist Perspective*. Lanham, MD: Rowman & Littlefield Publishers Inc., 2008.

Kheel, Marti. "An Ecofeminist Critique of Holistic Nature Ethics: Attending to Non-human Animals—Theodore Roosevelt, Aldo Leopold, Holmes Rolston III, Warwick Fox." PhD diss., Graduate Theological Union, California, 2000.

Kheel, Marti. "The Killing Game: An Ecofeminist Critique of Hunting." *Journal of the Philosophy of Sport,* Vol. 23, no. 1 (1996): 30–44.

Kheel, Marti. "The Liberation of Nature: A Circular Affair." *Environmental Ethics* 7, no. 2 (Summer 1985): 135-149.

"Killer Whale Still Carrying Dead Baby after 16 Days." BBC News, August 9, 2018. Killer whale still carrying dead baby after 16 days - BBC News.

Koehler, Matthew. "The Rocky Mountain Elk Foundation is Funding a $1000 'bounty' per Trapped and Killed Wolf in Idaho." Accessed January 28, 2019. The Rocky Mountain Elk Foundation is funding a $1000 'bounty' per trapped and killed wolf in Idaho | The Smokey Wire: National Forest News and Views (forestpolicypub.com).

Krause, Tom. *NTA Trapping Handbook—A Guide for Better Trapping*. National Trapper Association, 2008.

Krausman, Paul, and Shane P. Mahoney. "How the Boone and Crockett Club (B&C) Shaped North American Conservation," *International Journal of Environmental Studies,* Vol. 72 no. 5 (2015): 746–755, doi: https://doi.org/10.1080/00207233.2015.1073474.

Krug, Etienne G., Mercy, James A., Dahlberg, Linda L., and Anthony B. Zwi. *World Report on Violence and Health*. Geneva, World Health Organization, 2002.

Kuglin, Tom 2014. Famous Cow who Escaped Slaughter Moves to Helena Valley: Available at: Famous cow who escaped slaughter moves to Helena Valley | Local | helenair.com.

Kuglin, Tom and Seaborn Larson. Montana agencies hope new pheasant program benefits inmates, hunters | 406 Politics | helenair.com.

Kulowiec, Thomas G. et al. "2018 Mourning Dove Population and Research Status Report." 2018 Mourning Dove Population and Research Status Report.

Kymlicka, Will and Sue Donaldson. "Animal Rights, Multiculturalism and the Left." *Journal of Social Philosophy*, Vol. 45, no. 1 (2014): 116-135. (PDF) Animal Rights, Multiculturalism and the Left (2014) | Will Kymlicka and Sue Donaldson - Academia.edu.

Laundré, John, April 2020: article_are_hunters_conservationists_Laundre.pdf (predatordefense.org).

Lavigne, David. "Wildlife Conservation and the Pursuit of Ecological Sustainability: A Brief Introduction." In *Gaining Ground: In Pursuit of Ecological Sustainability*, edited by David Lavigne. Guelph, Canada: International Fund for Animal Welfare (IFAW), (2006): 1-18.

Leopold, Aldo. "Thinking Like a Mountain." *A Sand County Almanac and Sketches Here and There*. New York: Oxford Press, 1968.

Leopold, Aldo. "Thinking Like a Mountain." *A Sand County Almanac and Sketches Here and There*. London: Oxford Press, 1949.

Leopold, Aldo. *Game Management*. New York, NY: Charles Scribner's Sons, 1933.

Leopold, Bruce D., and Paul R. Krausman. "Science in Print: Lifeblood of the Society—Key Papers That Have Advanced the Wildlife Profession." *The Wildlife Professional*. Vol. 6 no. 3 (2012): 48–53.

Leopold, Bruce D. "Understanding Our Roots: The North American Wildlife Conservation Model." *The Wildlife Professional*, Vol. 4 no.3 (2010): 12.

Levy, Sharon. "Do Animals Suffer From Post-Traumatic Stress?" August 29, 2021. BBC News/Future. Do animals suffer from post-traumatic stress? - BBC Future.

Linzey, Andrew. *The Link Between Animal Abuse and Human Violence*. Brighton, UK: Sussex University Press, 2009.

Low, Philip. *The Cambridge Declaration on Consciousness in Non-Human Animals*, edited by Jaak Panksepp, Diana Reiss, David Edelman, Bruno Van Swinderen, and Christoph Koch. 2012. Retrieved from http://fcmconference.org/img/CambridgeDeclarationOnConsciousness.pdf.

Ludders, John W., Schmidt, Robert H., Dein, Joshua, and Patrice Klein. "Drowning is Not Euthanasia." *Wildlife Society Bulletin*, Vol. 27 no.3 (1999): 666–670.

Luke, Brian. "A Critical Analysis of Hunters' Ethics," *Environmental Ethics*, Vol. 19 no. 1 (1997): 27–28.

Luke, Brian. *Brutal: Manhood and the Exploitation of Animals.* University of Illinois Press, Urbana and Chicago, 2007.

Luke, Brian. "Taming Ourselves or Going Feral? Toward a Nonpatriarchal Metaethic of Animal Liberation." In *Animals and Women: Feminist Theoretical Explorations,* edited by Carol J. Adams and Josephine Donovan, 290 -319. Durham and London: Duke University Press, 1995.

Lusseau, David, and M.E.J. Newman. "Identifying the Role that Animals Play in Their Social Networks." *Proceedings of the Royal Society B: Biological Sciences*, Vol. 271 (Suppl. 6, 2004): S477–S481. doi:10.1098/rsbl.2004.0225.

Lutey, Tom. "After two brutal winters, stock growers press feds to expand livestock loss compensation." March 31, 2019, After two brutal winters, stock growers press feds to expand livestock loss compensation | State & Regional | billingsgazette.com.

Lynn, William. "Between Science and Ethics: What Science and the Scientific Method Can and Cannot Contribute to Conservation and Sustainability." *Gaining Ground: In Pursuit of Ecological Sustainability*, edited by David Lavigne. Guelph, Canada: International Fund for Animal Welfare (IFAW), 2006.

Macfarlene, Gary and Tim Preso. State-Sponsored Wolf Killing Ends in Idaho Faced with Looming Court Challenge, Idaho halts Unprecedented Program (January 2014). Accessed May 1, 2020. State-Sponsored Wolf Killing Ends in Idaho: (wildernesswatch.org).

MacKinnon, Catharine. "Of Mice and Men: A Feminist Fragment on Animal Rights." In *The Feminist Care Tradition in Animal Ethics: A Reader*, edited by Josephine Donovan and Carol Adams. New York, NY: Columbia University Press, 2007: 316-332.

Mahoney, Shane, and Valerius Geist. "The North American Model for Wildlife Conservation: Setting the Stage for Evaluation." *The North American Model of Wildlife Conservation*, edited by Shane Mahoney and Valerius Geist, 1-8. Baltimore: Johns Hopkins University Press, 2019.

Mahoney, Shane. "Hunting is by far the largest funding mechanism for wildlife, and the economic benefits go far beyond conservation." Accessed February 2,

2022. Hunting is by far the largest funding me | conservation-force (conservationforce.org).

Mahoney, Shane, and John Jackson III. "Enshrining Hunting as a Foundation for Conservation—The North American Model." *International Journal of Environmental Studies,* Vol. 70 no. 3 (2013): 448-459.

Mallory, Chaone. "Acts of Objectification and the Repudiation of Dominance: Leopold, Ecofeminism, and the Ecological Narrative." *Ethics and The Environment,* Vol. 6 no. 2 (2001): 59–89.

Manfredo, Michael J. *Who Cares About Wildlife? Social Science Concepts for Exploring Human-Wildlife Relationships and Conservation Issues.* New York, NY: Springer, 2008.

Manfredo, Michael J., Tara Teel, Leeann Sullivan, and Alia M. Dietsch. "Values, Trust, and Cultural Backlash in Conservation Governance: The Case of Wildlife Management in the United States." *Biological Conservation,* Vol. 214 (October 2017): 303-311.

Marino, Lori. "A Trans-Species Perspective on Nature." *On the Human.* 2010. Retrieved from https://nationalhumanitiescenter.org/on-the-human/2010/11/trans-species-perspective/.

McIntire, Mike, "Selling a New Generation on Guns," *The New York Times,* January 26, 2013, Selling a New Generation on Guns - The New York Times (nytimes.com).

Mcshane, Katie. "Anthropocentrism vs. Nonanthropocentrism: Why Should We Care? May 2007 *Environmental Values* 16, no. 2 (May 2007): 169-186. DOI:https://doi.org/10.3197/096327107780474555.

McIntyre, Rick. *The Reign of Wolf 21: The Saga of Yellowstone's Legendary Druid Pack,* Vancouver, Berkeley: Greystone Books, 2020.

Meine, Curt. *Aldo Leopold: His Life and Work.* Madison, WI: University of Wisconsin Press, 1988.

Merchant, Carolyn. *The Death of Nature: Women, Ecology, and the Scientific Revolution.* New York: Harper and Row, 1980.

Midgley, Mary. *Beast and Man: The Roots of Human Nature.* Brighton: Harvester, 1979.

Miller, Darren A. "Hunting As a Wildlife Management Tool." Guest Editorial. *The Wildlife Professional,* Vol. 4 no. 3 (Fall 2010): 8.

Mills, L. Scott. *Conservation of Wildlife Populations: Demography, Genetics, and Management.* Malden, MA: Blackwell Publishing, 2007.

Moag, Jeff. "How Socialite Rosalie Edge Became the Conscience of American Conservation." Adventure Journal. Historical Badass. November 8, 2019.

How Socialite Rosalie Edge Became the Conscience of American Conservation (adventure-journal.com).

Montana Department of Fish, Wildlife and Parks (MDFWP). "2020 Black Bear Harvest." FWP: Harvest Estimates (mt.gov).

Montana Department of Fish, Wildlife and Parks (MDFWP). "Furbearer Trapping Regulations," 2020. 2020 Furbearer Final for web 12172020.pdf.

Montana Department of Livestock. "Aerial Hunting Permits." Aerial Hunting Permits (mt.gov).

Montgomery, Robert A., Madeline Carr, Charlie Booher, Abigail Pointer, Brendan Mitchell, Natalie Smith, Keegan Calnan, Georgina Montgomery, Mordecai Ogada, and Daniel Kramer. "Characteristics That Make Trophy Hunting of Giant Pandas Inconceivable." *Conservation Biology*, Vol. 34, No. 4, (2020): 915-924.

Muth, Robert M., Rodney R. Zwick, M.E. Mather, John Organ, John J. Daigle and Sandra A. Jonker. "Unnecessary Source of Pain and Suffering or Necessary Management Tool: Attitudes of Conservation Professionals Toward Outlawing Leghold Traps." *Wildlife Society Bulletin,* Vol. 34 no. 3, (2006): 706–715.

Muth, Robert M. and Wesley Jamison. "On the Destiny of Deer Camps and Duck Blinds: The Rise of the Animal Rights Movement and the Future of Wildlife Conservation." *Wildlife Society Bulletin* Vol. 28 no. 4 (Winter, 2000): 841–851.

National Public Radio (NPR), September 29, 2021: Wildlife Officials Take 23 Species Off The Endangered List Due To Extinction : NPR.

National Public Radio (NPR)."Eisenhower's Military-Industry Complex Warning," 50 Years Later: NPR.

National Public Radio, (NPR). "British Study: Lobsters Might Experience Feelings, Including Pain." Interview of Dr. Jonathan Birch by A. Martinez, November 30, 2021. British study: lobsters might experience feelings, including pain: NPR.

National Trappers Association. "About the NTA." www.nationaltrappers.com.

Nee, Sean. "The Great Chain of Being" *Nature*, Vol. 435 (May 25, 2005): 429. doi:10.1038/435429a.

Nelson, Michael, Jeremy T. Bruskotter, John A. Vucetich, and Guillaume Chapron. "Emotions and the Ethics of Consequence in Conservation Decisions: Lessons From Cecil the Lion," *Conservation Letters* 9, (July 1, 2016): 302-306.

Nelson, Michael P., John A. Vucetich, Paul C. Paquet, and Joseph K. Bump. "An Inadequate Construct? North American Model: What's Flawed, What's Missing, What's Needed." *The Wildlife Professional*, (Summer 2011): 58-60.

Newhouse, Sewell. *The Trapper's Guide: A Manual of Instructions.* Big Timber, MT: The Buckskin Press, 1865.

Newman M.E.J. Assortative mixing in networks. 2002. *Physical Review Letters, 89*, art. no.-208701.

Nie, Martin, Christopher Barns, Jonathan Haber, Julie Joly, Kenneth Pitt, and Sandra Zellmer, "Fish and Wildlife Management on Federal Lands: Debunking State Supremacy," *Environmental Law*, Vol. 47 no. 4 (2017): 797-932.

Nonhuman Rights Project. "Humans are not The Only Animals Entitled to Recognition and Protection of Their Fundamental Rights." Accessed January 15, 2022. Nonhuman Rights Project.

Northeast Furbearer Resources Technical Committee of the Northeast Section of The Wildlife Society (NEFRTC). (2015). *Trapping and furbearer management in North American wildlife conservation.* Retrieved on March 19, 2021 from http://files.dnr.state.mn.us/recreation/hunting/trapping/na_furbearer_mgt.pdf.

Noske, Barbara. *Humans and Other Animals,* London: Pluto Press, 1989.

NRA-ILA: National Rifle Association's Institute for Legislative Action. *The Pittman-Robertson Act: Friend of the Hunter and Hunted.* Retrieved October 10, 2021, from: https://www.nraila.org/articles/20010828/pittman-robertson-act-friend-of-the-hu-1.

Nurse, Angus and Tanya Wyatt. *Wildlife Criminology.* Bristol: Bristol University Press, 2021.

Oommen, Meera A., Rosie Cooney, Madhuri Ramesh, Michael Archer, Daniel Brockington, Bram Buscher, Robert Fletcher, Daniel J.D. Natusch, Abi T. Vanak, Grahame Webb, Kartik Shanker. "The Fatal Flaws of Compassionate Conservation." The Society for Conservation Biology, Vol. 33, no. 4, (August 2019): 784-787. https://doi.org/10.1111/cobi.13329.

Organ, John F., Valerius Geist, Shane P. Mahoney, Steven Williams, Paul Krausman, Gordon Batcheller, Thomas R. Decker, Robert Carmichael, Priya Nanjappa, Ronald Regan, Rodrigo Medellin, Ruben Cantu, Richard McCabe, Scott Craven, Gary M. Vecellio, and Daniel J. Decker. "The North American Model of Wildlife Conservation." *The Wildlife Society Technical Review.* Vol. 12, no. 04 (2012). Bethesda MD: The Wildlife Society and The Boone and Crockett Club.

Organ, John F. "Born in the Hands of Hunters." *The Wildlife Professional,* Vol. 4 no. 3 (2010): 23–27.

Organ, John F., Robert Muth, Jan E. Dizard, Scot Williamson, and Thomas A. Decker. "Fair Chase and Humane Treatment: Balancing the Ethics of Hunting and Trapping." *Transactions of the North American Wildlife and Natural Resources Conference* Vol. 63 (1998): 528–543.

Orion: The Hunter's Institute. "What Does It Mean To Be An Ethical Hunter?" Accessed September 9, 2021. Ethics | Orion (orionhunters.org).

Ott, Jennifer. "Ruining the Rivers in the Snake Country: The Hudson Bay Company's Fur Desert Policy." *Oregon Historical Quarterly,* Vol. 104 no. 2 (2003): 166-195.

Ox Hunting Ranch. "Red Stag Hunting." Red Stag Hunting | 18,000 Acres in Texas | 60+ Species | Ox Ranch (oxhuntingranch.com).

Panksepp, Jaak. "Affective Consciousness: Core Emotional Feelings in Animals and Humans." *Consciousness and Cognition,* Vol. 14, no. 1 (2005): 30-80. https://doi.org/10.1016/j.concog.2004.10.004.

Paquet, Paul C. and C. T. Darimont. "Wildlife Conservation and Animal Welfare: Two Sides of the Same Coin?" *Animal Welfare,* Vol. 19 no. 2 (2010): 177–190.

Paquet, Paul C. "Wolves and Humans at the Crossroads," Plenary Session of the International Wolf Symposium, October 10-13, 2013a, Duluth, Minnesota.

Paquet, Paul C. "Why Does it Matter?" Plenary Session of the International Wolf Symposium – Wolves and Humans at the Crossroads. Oct. 10-13, 2013b. Duluth, Minnesota. Microsoft Word - Paquet, P. Oct. 2013. Wolf Speech IWC- Why does it matter?.docx (projectcoyote.org).

Parker et al. Demographic and Economic Trends in Urban-Suburban and Rural Communities. Pew Research Center. May 22, 2018. https://www.pewsocialtrends.org/2018/05/22/demographic-and-economic-trends-in-urban-suburban-and-rural-communities/.

Perkins, Derrick. "Commissioners Back Feds in Lawsuit Over Wolverines," The Western News, 2021. www.thewesternnews.com.

Peterson, Christine. "25 years after returning to Yellowstone, wolves have helped stabilize the ecosystem: New research shows that by reducing populations and thinning out weak and sick animals, wolves have a role in creating resilient elk herds," July 10, 2020. 25 years after returning to Yellowstone, wolves have helped stabilize the ecosystem (nationalgeographic.com).

Peterson, M. Nils, and Michael Paul Nelson. "Why the North American Model of Wildlife Conservation is Problematic for Modern Wildlife Management." *Human Dimensions of Wildlife,* Vol 22. no. 1 (September 2016): 1-12.

Phillips, Allie. (2014). *Understanding the Link Between Violence to Animals and People: A Guidebook for Criminal Justice Professionals*. Retrieved on February 6, 2020, from: http://nationallinkcoalition.org/wp-content/uploads/2014/06/Allies-Link-Monograph-2014.pdf.

Plumwood, Val. "Has Democracy Failed Ecology? An Ecofeminist Perspective." In *Ecology and Democracy*, edited by Freya Matthews, 134-168. London: Frank Cass Ltd., 1996.

Plumwood, Val. "Nature, Self, and Gender: Feminism, Environmental Philosophy, and the Critique of Rationalism." *Hypatia Special Issue: Ecological Feminism*. Vol. 6 no. 1 (1991): 3–27.

Posewitz, Jim. "The Hunter's Ethic: The Past, the Peril, and the Future." *The Wildlife Professional* Vol. 4 no. 3 (2010): 32.

Poten, Connie. "A Shameful Harvest: America's Illegal Wildlife Trade." National Geographic Magazine, Vol. 180, no. 3. (September 1991): 106 – 132.

Predator Defense. "Bobcat Rescued From Illegal Strangling Snare." April, 2012. http://www.predatordefense.org/features/snares_bobcat.htm.

Professional Outdoor Media. "Support SCI President Paul Babaz for NRA Board of Directors," Member News, 2019. Professional Outdoor Media. www.professionaloutdoormedia.org.

Proulx, Gilbert. *Intolerable Cruelty: The Truth Behind Killing Neck Snares and Strychnine.* Canada: Alpha Wildlife Publications, 2018.

Proulx, Gilbert, Ryan K. Brook, Marc Cattet, Chris Darimond, and Paul C. Paquet. "Poisoning Wolves with Strychnine is Unacceptable in Experimental Studies and Conservation Programmes." *Environmental Conservation:* page 1 of 2 (2016). doi:10.1017/S0376892915000211.

Proulx, Gilbert and Dwight Rodtka. "Steel-Jawed Leghold Traps and Killing Neck Snares: Similar Injuries Command Change to Agreement on International Humane Trapping Standards," *Journal of Applied Animal Welfare Science*, 20, 2 (2017): 1-6. https://doi.org/10.1080/10888705.2017.1286989.

Proulx, Gilbert. "Review of Current Mammal Trap Technology in North America." In *Mammal Trapping,* edited by Gilbert Proulx, 1-46. Alberta, Canada: Alpha Wildlife Research & Management Ltd., 1999.

Raincoast Conservation Foundation. "Safeguarding Coastal Carnivores in the Southern Great Bear Rainforest tenure." Accessed December 5, 2021, Southern Great Bear Rainforest tenure | Raincoast Conservation Foundation.

Ramp, Daniel and Marc Bekoff. "Compassion as a Practical and Evolved Ethic for Conservation." *BioScience* 65 (2015): 323-327.

Realtree, "Female Shooting Sports Participation Continues Strong Growth," August 23, 2018. www.realtree.com.

Regan, Tom. *The Case for Animal Rights.* Berkeley and Los Angeles, CA: University of California Press, 1983.

Reggente et al. (2016). Nurturant Behavior Toward Dead Conspecifics in Free-ranging Mammals: New Records for Odontocetes and a General Review. *Journal of Mammalogy,* Volume 97, Issue 5, 27 (September 2016): 1428–1434. https://doi.org/10.1093/jmammal/gyw089.

Reiger, John F. *American Sportsmen and the Origin of Conservation,* 3rd ed. Corvallis, OR: Oregon State University Press, 2001.

Reiger, John F. *American Sportsmen and the Origin of Conservation.* Corvallis, OR, Oregon State University Press, 1975.

Responsive Management. 2015. "How to Talk to the Public About Hunting: Research Based Communication Strategies. http://www.responsivemanagement.com/download/RM_E News/RM_HuntPublic_newsletter.pdf. Accessed September 20, 2021.

Riffkin, Rebecca. "In U.S., More Say Animals Should Have Same Rights as People." *Gallup,* May 18, 2015. Retrieved November 2, 2021, from http://www.news.gallup.com/poll/183275/say-animals-rights-people.aspx.

Ripple, WJ, C. Wolf, TM Newsome, MG Betts, G Ceballos, F. Courchamp, MW Hayward, B Van Valkenburgh, AD Wallach, and B Worm. "Are We Eating the World's Megafauna to Extinction?" *Conservation Letters* 12(3), (January 1, 2019): 1-10. DOI: 10.1111/conl.12627

RMEF (The Rocky Mountain Elk Foundation). "RMEF Strongly Supports Delisting, State Management of Wolves," Conservation: News, October, 2020. Rocky Mountain Elk Foundation.www.rmef.org.

RMEF (The Rocky Mountain Elk Foundation). "RMEF Renews Support for Foundation of Wildlife Management." *Elk Network.* Rocky Mountain Elk Foundation Website, accessed 2017, Feb 1. from: RMEF Renews Support for Foundation for Wildlife Management | Rocky Mountain Elk Foundation.

RMEF (The Rocky Mountain Elk Foundation). "The North American Wildlife Conservation Model." *Hunting is Conservation.* Rocky Mountain Elk Foundation Website, accessed 2015, Dec 1. from: http://www.rmef.org/Conservation/HuntingIsConservation/NorthAmericanWildlifeConservationModel.aspx.

Rollin, Bernard E. *The Unheeded Cry: Animal Consciousness, Animal Pain, and Science.* Oxford: Oxford University Press, 1998.

Rutberg Allen T. "Why State Agencies Should Not Advocate Hunting or Trapping," *Human Dimensions of Wildlife* 6, no. 1 (January 2001): 33-37. DOI:https://doi.org/10.1080/10871200152668670.

Safina, Carl. *Beyond Words: What Animals Think and Feel.* New York, NY: Henry Holt and Company, LLC., 2015.

Sax, Joseph L. "The Public Trust Doctrine in Natural Resource Law: Effective Judicial Intervention." *Michigan Law Review* Vol. 68, (1970): 471–566.

Sax, Joseph L. "Introduction to the Public Trust Doctrine." In *The Public Trust Doctrine and Its Application to Protecting Instream Flows: The Proceedings of a Workshop (NIFPA-08),* edited by G.E. Smith and A.R. Hoar, 5–12. Anchorage, AK: Alaska Department of Fish and Game and U.S. Fish and Wildlife Service-Region 7, 1999.

Science Daily. "Stark warning: Combating ecosystem collapse from the tropics to the Antarctic." February 26, 2021. Stark warning: Combating ecosystem collapse from the tropics to the Antarctic %2D%2D ScienceDaily.

Scott, Tristan. "Hunters and Trappers Call for Fewer Wolves: Sportsmen and Lawmakers aim for a More Liberal Wolf-hunting Season in an Effort to Draw Down the Population." February 1, 2019. Hunters and Trappers Call for Fewer Wolves - Flathead Beacon.

Seamans, M. E. Mourning Dove Population Status, 2018. U.S. Department of the Interior, Fish and Wildlife Service, Division of Migratory Bird Management, Washington, D.C.

Seegert, Alf. "Pixels and Pathos: Video Games and Empathy." Presented at: Interdisciplinary Symposium on Empathy, Contemplative Practice and Pedagogy, the Humanities, and the Sciences. University of Utah, USA. Pixels_and_Pathos_Video_Games_and_Empath.pdf.

Serfass, Thomas L., Robert Brooks, and Jeremy T. Bruskotter. "North American Model of Wildlife Conservation: Empowerment and Exclusivity Hinder Advances in Wildlife Conservation." *Canadian Wildlife Biology and Management.* Vol. 7, no. 2. (2018): 102-118.

Servheen, Christopher (on behalf of many scientists with experience in wildlife and public land management), "Scientists Say Gianforte's Anti-Wolf, Anti-Grizzly Policies in Montana Have No Scientific Basis: Prominent Group of Wildlife Professionals With 1,500 Years of Experience Condemn Montana's New Laws Targeting Wolves. Already Pups From Popular Yellowstone Wolf Pack Have Been Killed," October 2, 2021. Scientists Say Gianforte's Anti-Wolf, Anti-Grizzly Policies In Montana Have No Scientific Basis (mountain-journal.org).

Singer, Peter. *Animal Liberation.* New York: Avon Books, Inc., 1991.

Singer, Peter. *Animal Liberation: A New Ethics For Our Treatment of Animals*. New York: Avon Books, 1st edition, 1977.

Slicer, Deborah. "More Joy." *Ethics and the Environment*, Vol. 20, no., 2, (Fall 2015): 1-23.

Slicer, Deborah. "Your Daughter or Your Dog?" *Hypatia. Special Issue: Ecological Feminism*, Vol. 6 no. 1 (Spring, 1991): 108–121.

Smith, Marc and Donald Molde. 2015. Microsoft Word - Smith Molde Wildlife Funding spreadsheet updated June15.doc (nrwm.org).

Sorensen, Eric. "Research Finds Lethal Wolf Control Backfires on Livestock," *WSU Insider*. (December 3, 2014). https://news.wsu.edu/2014/12/03/research-finds-lethal-wolf-control-backfires-on-livestock/.

Spanjol, Kimberly. "Wild Crimes: The Exploitation of our World's Most Vulnerable Animals," webinar, Wolf Conservation Center, Wild Crimes: The Exploitation of our World's Most Vulnerable Animals | Wolf Conservation Center (nywolf.org).

Spencer, Daniel T. *Gay and Gaia: Ethics, Ecology, and the Erotic*. Cleveland: The Pilgrim Press, 1996.

Stalling, Dave. "A Once-Proud Conservation Group Has Lost its Way." Opinion. High Country News (August 31, 2012). Retrieved from https://www.hcn.org/wotr/a-once-proud-conservation-group-has-lost-its-way.

Stallwood, Kim. "Are We Smart Enough to Know When to Take the Political Turn for Animals?" In: *Ethical and Political Approaches to Nonhuman Animal Issues, edited by* Andrew Woodhall and Gabriel Garmendia da Trindade, 275-303. Switzerland: Springer/Palgrave Macmillan Cham, 2017. https://doi.org/https://doi.org/10.1007/978-3-319-54549-3.

Stastny, Leonard. "Trapping: Steel Traps Should Never Be Allowed." (2012) http://missoulian.com/news/opinion/mailbag/trapping-steel-traps-should-never-be-allowed/article_f2a9f646-37ac-11e1-8e71-001871e3ce6c.html.

Stout, M. (2005). *The Sociopath Next Door*. New York, NY: Fine Communications.

Sullivan's Trap Line Bulletin Board. Fur Handling Archives. "Carcass Disposal." http://sullivansline.com/forums/ubbthreads.php?ubb=showflat&Number=11593&page=all http://newwest.net/main/article/a_montana_trapper_confesses/.

Sussman, Robert. *Man the Hunted: Primates, Predators and Human Evolution*. 2006. See information on this book at: 'Man the Hunter' theory is debunked in new book - The Source - Washington University in St. Louis (wustl.edu).

Swan Mountain Outfitters. "Montana Black Bear Hunts." Accessed December 20, 2021. Montana Black Bear Hunts | Spring Bear Hunts (swanmountainoutfitters.com).

"Eight wolves were found poisoned in Oregon. Police are asking for the public's help: Conservation groups call the deaths 'shocking and heartbreaking' and have offered a $43,000 reward for information." The Guardian, 2021. Eight wolves were found poisoned in Oregon. Police are asking for the public's help | Oregon | The Guardian.

Taylor, Paul. *Respect for Nature: A Theory of Environmental Ethics*. Princeton: Princeton University Press, 1986.

Teel, Tara, and Michael J. Manfredo. "Understanding the Diversity of Public Interests in Wildlife Conservation." *Conservation Biology*, Vol. 24 no.1, (February, 2009): 128-139.

The Institute for Critical Animal Studies. "About." About | Institute for Critical Animal Studies (ICAS).

The Wildlife Society, "TWS Issue Statement: Baiting and Supplemental Feeding of Game Wildlife Species." Last modified March 24, 2020. TWS Issue Statement: Baiting and Supplemental Feeding of Game Wildlife Species - The Wildlife Society.

The Wildlife Society, "TWS Position Statement: Firearms." Last modified March 23, 2020. TWS Position Statement: Firearms - The Wildlife Society.

The Wildlife Society. TWS Standing Position: Trapping Furbearers." *The Wildlife Society*. Bethesda, MA: The Wildlife Society, July 7, 2020a. TWS Standing Position: Trapping Furbearers - The Wildlife Society.

The Wildlife Society. Final TWS Position Statement: "The North American Model of Conservation," Bethesda, MA: The Wildlife Society, 2020b. http://joomla.wildlife.org/documents/positionstatements/41NAModel%20Position%20Statementfinal.pdf.

Total Outdoor Adventures. "Spring vs. Fall Black Bear Hunts." February 28, 2019. Spring vs. Fall Black Bear HuntsTotal Outdoor Adventures | Total Outdoor Adventures (toaltd.com).

Training Institute for the Prevention of Strangulation. (strangulationtraininginstitute.com).

Trauger, D.L. and P.L. Kennedy. "Pioneering Professionals: Sharing a Passion for Wildlife Conservation." *The Wildlife Professional* Vol. 6 no. 3 (2012): 32.

Trefethen, James B. *An American Crusade for Wildlife*. New York: Boone & Crockett Club, Winchester Press, 1975.

Trek International Safaris. "Wing Shooting Destinations." Accessed December 5, 2021. Wing Shooting Destinations - Trek Safaris.

Treves, Adrian, Guillaume Chapron, Jose V. Lopez-Bao, Chase Shoemaker, Apollonia R. Goeckner and Jeremy T. Bruskotter. "Predators and the Public

Trust." *Biological Reviews*, no. 92, (2017): 248–270. doi: https://doi.org/10.1111/brv.12227.

Twine, Richard. "Revealing the 'Animal-Industrial Complex' – A Concept & Method for Critical Animal Studies?" *Journal for Critical Animal Studies*, Vol. 10, no. 1, (2012): 12-39. (ISSN1948-352X).

United States Department of the Interior (DOI), "Everything you need to know about hunting on public lands," last modified September 1, 2017, "Everything you need to know about hunting on public lands" | U.S. Department of the Interior (doi.gov)

United States Department of the Interior, U.S. Fish and Wildlife Service, and U.S. Department of Commerce, U.S. Census Bureau. 2016 National Survey of Fishing, Hunting, and Wildlife-Associated Recreation. fhw16-nat.pdf (census.gov).

Unites States Fish and Wildlife Service (USFWS), Migratory Bird Program. "Adaptive Harvest Management." Accessed December 17, 2021. https://www.fws.gov/birds/management/adaptive-harvest-management.php.

United States Fish and Wildlife Service (USFWS). "Hunters as Conservationists." Hunters as Conservationists | U.S. Fish and Wildlife Service (fws.gov).

Van de Pitte, Margaret. "The Moral Basis for Public Policy Encouraging Sport Hunting." *Journal of Social Philosophy*, Vol. 34 no. 2 (2003): 256–266.

Van Dyke, Fred. "The History and Distinctions in Conservation Biology." In *Conservation Biology: Foundations, Concepts, Applications*. Wheaton College, IL: Springer, 2008. Accessed Sept. 9, 2013: E-Book.

Vucetich, John A., and Michael P. Nelson. "The Infirm Ethical Foundations of Conservation." In: *Ignoring Nature No More: The Case for Compassionate Conservation,* edited by Marc Bekoff, 1-25. Chicago, IL: University of Chicago Press, 2013.

Vucetich, John A, and Rolf Peterson. "Using Basic Principles of Wildlife Management to Evaluate the Prospects for a Public Wolf Harvest in Michigan." Testimony presented to Michigan's Natural Resource Hearings. Accessed February 1, 2022: Retrieved from: http://www.thewildlifenews.com/wp-content/uploads/2013/05/Evaluation-MI-wolf-harvest-Vucetich--Peterson-1May2013.pdf.

Walker, Margaret. 1989. Moral understandings: Alternative "epistemology" for a feminist ethics. *Hypatia* 4(2): 15–28.

Wallach, Arian D., Chelsea Batavia, Marc Bekoff, Shelley Alexander, Liv Baker, Dror Ben-Ami, Louise Boronyak, Adam P. A. Cardilin, Yohay Carmel, Danielle Celermajer, Simon Coghlan, Yara Dahdal, Jonatan J. Gomez, Gisela Kaplan,

Oded Keynan, Anton Khalilieh, Helen Kopnina, William S. Lynn, Yamini Narayanan, Sophie Riley, Francisco J. Santiago-Ávila, Esty Yanco, Miriam A. Zemanova, and Daniel Ramp. "Recognizing Animal Personhood in Compassionate Conservation." *Conservation Biology*, Volume 34, no. 5, (2020): 1097–1106. Society for Conservation Biology DOI: 10.1111/cobi.13494.

Wallach, Arian D., Bekoff, Marc, Batavia, Chelsea, Nelson, Michael P. and Daniel Ramp. "Summoning Compassion to Address the Challenges of Conservation." *Society for Conservation Biology*, Vol. 0, No. 0, 2018: 1–11.

Warren, Karen J. "The Power and Promise of Ecological Feminism." In *Environmental Ethics*, Vol. 12, no. 2 Summer (1990): 125-146. doi: https://doi.org/10.5840/enviroethics199012221.

White, Bryant, Thomas Decker, Michael J. O'Brien, John F. Organ, and Nathan M. Roberts. "Trapping and Furbearer Management in North American Wildlife Conservation." *International Journal of Environmental Studies*, Vol. 72 no. 5 (September 2015): 756–769.

White, Bryant, Clifford Brown, and Thomas Decker. "New Guidelines for Furbearer Trapping: Science Improves an Age-Old Pursuit." *The Wildlife Professional.* (Fall 2010): 66–71.

White, Gary C. and Chad J. Bishop. "A Bountiful Harvest for Science: How Conservation Science Benefits from the Study of Game Species." *The Wildlife Professional,* (Fall 2010): 42–46.

White, Lynn, Jr. "The Historical Roots of Our Ecological Crisis." *Science* Vol. 155, No. 3767 (1967): 1203-1207.

Wielgus Robert, and Kaylie Peebles. "Effects of Wolf Mortality on Livestock Depredations." *PLOS One* Vol. 9 no. 12 (Dec. 3, 2014): e113505. DOI: https://doi.org/10.1371/journal.pone.0113505

Wildlife For All. "A National Campaign to Reform State Wildlife management to be More Ecologically, Driven, Democratic and Compassionate." Accessed February 2, 2022. Wildlife For All is a national campaign to reform state wildlife management.

Wilkinson, Todd. "Montana Defiantly Puts Yellowstone Wolves in its Crosshairs," September 9, 2021. Montana Defiantly Puts Yellowstone Wolves In Its Crosshairs (mountainjournal.org).

Wilkinson, Todd. "A Death Of Ethics: Is Hunting Destroying Itself?: From Killing Baboon Families to Staging Predator-Killing Contests, Hunters Stand Accused of Violating the North American Model of Wildlife Conservation. Now They're Being Called Out by Their Own." December 12, 2018. A Death Of Ethics: Is Hunting Destroying Itself? (mountainjournal.org).

Wise, Steven. "Legal Personhood and the Nonhuman Rights Project." *Animal Law Review,* 17 (2010):1-12. lralvol17_1_1.pdf (animallaw.info).

Wolfe, Gary J., Vermillion Dan, Aldrich Tim, Colton Shane, Doherty Steve, Tortlotte Matt, and Logan Brower. "Montana Wolf Policies are Destroying State's Reputation as Beacon for Wildlife Management: Seven Respected Former Wildlife Commissioners, All Hunters, Condemn Montana Governor and Lawmakers for Their Callous, Unscientific, Promotion of Wolf Slaughter," September 26, 2021. https://mountainjournal.org/respected-former-wildlife-commissioners-condemn-montana-slaughter-of-wolves.

Wolfe, M. L. and J.A. Chapman. "Principles of Furbearer Management." In: *Wild Furbearer Management and Conservation in North America,* edited by Milan Novak, James A. Baker, Martyn Obbard, and Bruce Malloch, 101-102. Toronto, Ontario: Ontario Ministry of Natural Resources, 1986.

Wood, Julia T. "Feminist Standpoint Theory." *Encyclopedia of Communication Theory*. Ed. Thousand Oaks, CA: SAGE 2009, 397-99. SAGE Reference Online. Web. 29 Jun. 2012.

Work That Reconnects. "Council of All Beings." Work That Reconnects Network. Last modified 2017. Council of All Beings (workthatreconnects.org).

World Health Organization. "Executive Summary." *Global Status Report on Violence Prevention, 2014.* http://apps.who.int/iris/bitstream/10665/145087/3/WHO_NMH_NVI_14.2_eng.pdf?ua=1&ua=1.

World Wide Fund for Nature (WWF). "Why Are We Loosing Nature?" Accessed January 2022. Living Planet Report | LPR | Why are we losing nature? (panda.org).

World Wide Fund for Nature (WWF). "Living Index 2020." Living Planet Report | LPR | Why are we losing nature? (panda.org).

Wuerthner, George. "North American Model of Wildlife Conservation And Wolves: Wolf Killing Makes Mockery of North American Model of Wildlife Conservation." *The Wildlife News*. October 14, 2013. thewildlifenews.com/2013/10/14/north-american-model-of-wildlife-conservation-and-wolves/.

Yeoman, Barry. "When Animals Grieve: Scientists are uncovering evidence that humans are not the only animals that mourn their dead." January 30, 2018. When Animals Grieve (nwf.org).

Young Ashley, Kathayoon K. Khalil, and Jim Wharton. "Empathy for Animals: A Review of the Existing Literature.". Curator The Museum Journal Vol. 61, no. 1, (March 2018): 327-343. DOI:https://doi.org/10.1111/cura.12257.

Zeckhauser, Richard. "Human Hunters and Nonhuman Predators: Fundamental Differences." *Proceedings of the National Academy of Sciences*, 114, no. 1 (January 2017): 13–14, www.pnas.org/cgi/doi/10.1073/pnas.1617003114.

Index[1]

[1] Note: Page numbers followed by 'n' refer to notes.

A. Heister, *Beyond the North American Model of Wildlife Conservation*, The Palgrave Macmillan Animal Ethics Series, https://doi.org/10.1007/978-3-031-14149-2

The manufacturer's authorised representative in the EU is Springer Nature Customer Service Centre GmbH, Europaplatz 3, 69115 Heidelberg, Germany. If you have any concerns regarding our products, please contact ProductSafety@springernature.com

Printed and bound by CPI Group (UK) Ltd, Croydon, CR0 4YY
28/09/2026
02224072-0010